A Series In

The History of Modern Physics, 1800-1950

The History of Modern Physics, 1800-1950

TITLES IN SERIES

INTRODUCTORY NOTE

The Tomash series in the History of Modern Physics offers the opportunity to follow the evolution of physics from its classical period in the nineteenth century when it emerged as a distinct discipline, through the early decades of the twentieth century when its modern roots were established, into the middle years of this century when physicists continued to develop extraordinary theories and techniques. The one hundred and fifty years covered by the series, 1800 to 1950, were crucial to all mankind not only because profound evolutionary advances occurred but also because some of these led to such applications as the release of nuclear energy. Our primary intent has been to choose a collection of historically important literature which would make this most significant period readily accessible.

We believe that the history of physics is more than just the narrative of the development of theoretical concepts and experimental results: it is also about the physicists individually and as a group—how they pursued their separate tasks, their means of support and avenues of communication, and how they interacted with other elements of their contemporary society. To express these interwoven themes we have identified and selected four types of works: reprints of "classics" no longer readily available; original monographs and works of primary scholarship, some previously only privately circulated, which warrant wider distribution; anthologies of important articles here collected in one place; and dissertations, recently written, revised, and enhanced. Each book is prefaced by an introductory essay written by an acknowledged scholar, which, by placing the material in its historical context, makes the volume more valuable as a reference work.

The books in the series are all noteworthy additions to the literature of the history of physics. They have been selected for their merit, distinction, and uniqueness. We believe that they will be of interest not only to the advanced scholar in the history of physics, but to a much broader, less specialized group of readers who may wish to understand a science that has become a central force in society and an integral part of our twentieth-century culture. Taken in its entirety, the series will bring to the reader a comprehensive picture of this major discipline not readily achieved in any one work. Taken individually, the works selected will surely be enjoyed and valued in themselves.

A Series In

The History of Modern Physics, 1800-1950

VOLUME III

American Physics in Transition

American Physics in Transition

A HISTORY OF
CONCEPTUAL CHANGE IN THE
LATE NINETEENTH CENTURY

by Albert E. Moyer

Tomash Publishers

LOS ANGELES / SAN FRANCISCO

ACKNOWLEDGMENTS

For helping me to prepare this study, I thank Daniel Siegel, Ronald Numbers, Victor Hilts, David Lindberg, Paul Conkin, Richard Hirsh, and Lynette Moyer. I also am indebted to Daniel Kevles, not only for suggesting in his writings various of the themes I have investigated, but also for identifying through statistical techniques, some of the lesser-known yet productive physicists whose work I have analyzed. Finally, I thank Gerald Holton and Katherine Sopka as well as Adele Clark and Erwin Tomash for their editorial guidance.

A.E.M.
June 1982

SECOND PRINTING, 1986

This printing is part of a copublishing agreement between Tomash Publishers and the American Institute of Physics.

Library of Congress Cataloging in Publication Data

Moyer, Albert E., 1945–
    American physics in transition.

    (History of modern physics, 1800–1950; v. 3)
    Bibliography: p.
    Includes index.
    1. Physics—History—United States. I. Title.
II. Series.
QC9.U5M68   1983   530'.0973   82-50750
ISBN 0-938228-06-4

For June and Bob Moyer

CONTENTS

CONCEPTUAL FERMENT AND
REVOLUTIONARY EXPECTATIONS

Foreword

MOMENTOUS upheavals, shattering events—in a word, revolutions—attract the attention of the historian, and historians of science have scientific revolutions to study. The classic scientific revolution, object of intensive historical study past and present, is, of course, the scientific revolution of the seventeenth century, which saw the replacement of the ancient and medieval scientific tradition by a new scientific outlook, in which the science of mechanics, as articulated especially by Isaac Newton, played a central role. The Newtonian, mechanical worldview flourished and bore fruit over a period of two hundred years, but it also was ultimately to be superseded: in what has been called the second scientific revolution, or the scientific revolution of the twentieth century (ca. 1895–1930), the mechanical worldview gave way to a radically different vision of scientific content and procedure, associated with the theory of relativity, quantum theory, and atomic physics. Both the first and second scientific revolutions affected primarily the foundations of the physical sciences, but both ultimately had profound implications for the biological sciences as well. Both of these scientific revolutions, then, were fundamental and thoroughgoing, affecting the whole of the corpus of science, and they are both therefore highly worthy of investigation. According to the dictum, however, that "history must be past and the more past the better . . . for its character as history,"* the first of these scientific revolutions has had much more attention lavished on it by historians than the second. There is great need, then, for the kind of historical research that will form the basis for a historiography of the second scientific revolution comparable in depth and sophistication to that we already have for the first. It is toward that general goal that the present work is oriented.*

*Thomas Mann, *Der Zauberberg* (1924, reprint Berlin: G.B. Fischer, 1958), p. 3, my own somewhat free translation.

More specifically, Albert Moyer's research addresses a problem that is central in the historiographies of both the first and second scientific revolutions. In the decades immediately following each of these revolutions, an extensive popular literature appeared, singing the praises of the revolution just past and bringing its message to a broad audience. Common in this self-congratulatory literature was the tendency to draw a strong contrast between the enlightened and dynamic scientific enterprise fostered by the revolution, and the benighted, wrongheaded, and stagnant system of belief supposed to have subsisted in prerevolutionary times. Historical accounts were then based on this literature, and these tended to overemphasize elements of discontinuous change and unprecedented novelty in the revolution, while giving short shrift to elements of continuity and precedent. This kind of history has had to be rewritten: looking back with more care and detachment, later historians have been able to find much more continuity, many more adumbrations of things to come in the prerevolutionary periods. This task is already far advanced with respect to the medieval and renaissance precedents for the scientific revolution of the seventeenth century; as concerns nineteenth-century roots of the novelties of twentieth-century science, our understanding is as yet much more rudimentary, and Albert Moyer's excellent work in this area thus constitutes a most welcome and important contribution.

In particular, Moyer has chosen to study late nineteenth-century attitudes toward the foundations of the physical sciences as reflected in the writings of American physicists. A growing dissatisfaction with the mechanical foundations of the physical sciences, and an associated willingness to search for alternatives, has already been documented for certain European scientists during that period; what we now need to know is how widespread these attitudes were, both in Europe and abroad. Among the various national physics communities outside of Europe, the American was the most vigorous at the time, and careful study of it is therefore most important for our understanding of the broader diffusion (and also independent development) of the new ideas. Beyond this, the scientific orientations of American physicists during this crucial period are important for our understanding of the development of the American physics community itself. This community emerged in the early twentieth century as an institutionally mature and intellectually active community of world significance; an appreciation of the intellectual preparedness with which this community greeted the scientific revolution of the twentieth century helps us to understand the phenomenon of its subsequent rapid rise to world significance and then preeminence. For the insight it gives into the development of American physics and American science, then, as well as for the broader understanding of the roots of the second scientific revolution that it furnishes, *American Physics in Transition: A History of Conceptual Change in the Late Nineteenth Century* constitutes a most significant addition to the literature of the history of science.

DANIEL M. SIEGEL
University of Wisconsin—Madison

Introduction: *Ideas and Institutions*

MAX PLANCK's quantum hypothesis and Albert Einstein's special theory of relativity helped trigger one of the major conceptual upheavals in the history of physics. The international community of physicists first perceived the full magnitude of the disruption between about 1905 and 1915, a few years after Planck's initial breakthrough in 1900 and Einstein's in 1905. The upheaval persisted into subsequent years, culminating in Einstein's general theory of relativity and in the quantum mechanics of Niels Bohr, Werner Heisenberg, and others. In its broadest outlines, this twentieth-century revolution in physics meant the subordination of traditional mechanical outlooks. By the late 1920s, there remained only conditional support for explanations of physical events in terms of processes that obeyed the laws of classical mechanics.

In contrast with the decades directly following 1905, the preceding thirty years or so in the world of physics might appear to have been a period of inactivity and quiet. This seems especially true for the United States, a country commonly considered to have been lacking in institutional support for physics and only marginally in touch with the major developments of international physics. If Europe, the locus of the eventual revolution, was in a period of calm, surely the fledgling United States was even more placid in the years from about 1870 to 1905. But was Europe, let alone the United States, in a state of prerevolutionary tranquillity?

As historians and scientists generally know, portentous events were in fact occurring in European physics during the decade preceding 1905. Most obvious, the discovery of X rays and radioactivity in the late 1890s jarred the scientific community. Moreover, as specialists in the history of modern physics have shown, there were changes of a subtler nature beginning even prior to 1895. Russell McCormmach, Martin Klein, Gerald Holton, Thomas Kuhn, and Stephen Brush have variously established that German, French, British, and Dutch physicists were questioning traditional mechanical perspectives during the final decades of the nineteenth century.

[xvii]

Increasingly, these Europeans were opting for recent electromagnetic, thermodynamic, and statistical views of nature as well as operational and skeptical outlooks on scientific inquiry. The result, according to these historians, was "turmoil," "crisis," "ferment," and "revolutionary expectations"—all prior to the impact of Planck and Einstein. P. M. Harman has reached a similar conclusion in his more comprehensive survey of nineteenth-century physics. In viewing the "revolution" associated with relativity and quantum theory from the perspective of the preceding century, he has been led to emphasize "the continuity of ideas that is present even in episodes of striking conceptual change."[1]

Did these ideas and signs of a probable impending upheaval extend across the Atlantic to the United States? Yes. There was conceptual ferment among Americans which paralleled that among Europeans. Physics in the United States indeed reflected both the earlier, subtler unrest and the later, more pronounced tumult that foretold a probable upheaval. Borrowing from Henry Adams's description of the cultural transitions that were occurring around 1900, we can characterize the intellectual trend in American physics during the period from about 1870 to 1905 as being from "unity" to "multiplicity."[2] Slowly at first, but more rapidly around 1895, American physicists shifted from the unity hoped for but never achieved in the traditional mechanical view of nature, to the multiplicity associated with the emerging, but as yet unsettled, scientific patterns of the dawning twentieth century.

What makes this period of conceptual transition in American physics historiographically distinctive is that it coincided with the exact period during which the American community of physicists first approached institutional maturity and achieved a sense of professional identity. Historians of science, particularly Daniel Kevles, have traced a variety of such institutional and professional developments. During the late nineteenth century, for example, American physicists for the first time obtained their own organizations and journal: Section B–Physics was established within the American Association for the Advancement of Science in 1882; the American Physical Society was founded in 1899; and the *Physical Review* was first published in 1893. In addition, Americans now had, within their own country, universities that offered both doctoral degrees and research opportunities in physics: The Johns Hopkins University was opened in 1876; Harvard's Jefferson Physical Laboratory was set afoot in 1884; and Chicago's Ryerson Laboratory was dedicated in 1894. Finally, American physicists began to have access to well-equipped laboratories funded by government, private foundations, and industry: both the National Bureau of Standards and the Carnegie Institution were organized in 1901; and the laboratories at American Telephone and Telegraph and at General Electric were moving toward basic research soon after 1900.

Kevles has further established, in a numerical analysis of the American physics community around the turn of the century, that the number of publishing physicists surged dramatically, from an aggregate of about 240 in the years 1870–1893 to about 650 in 1894–1915. In addition, more and more American physicists were earning doctoral degrees. Not only did these degrees come increasingly from American rather than European universities, but they came from an expanding range of American universities. Of the mere 33 domestic doctorates in physics awarded during 1870–1893, about 80% were from just 3 Eastern schools—Johns Hopkins, Yale, and Harvard. But of the 310 American doctorates awarded during 1894–1915, about 70% were from 7 universities—Cornell, Hopkins, Chicago, Harvard, Columbia, Princeton, and Yale—while the remaining 30% were from at least 15 other schools across the nation.[3] Given such figures, it is not surprising to learn from historians of science Paul Forman, John Heilbron, and Spencer Weart, that by the turn of the century, American academic physics was, for the first time, on a par with academic physics of France, Germany, and England in level of expenditures, volume of publications, and number of practitioners.[4]

The coincidence of this move toward institutional maturity and the shift toward new scientific outlooks makes the study of the latter all the more interesting. This temporal overlap suggests that the American physics community during its late adolescence and early maturity was being imprinted with the distinctive patterns of emerging twentieth-century thought. Perhaps these fresh roots help explain why American physicists in later decades so vigorously embraced the quantum theory and the theory of relativity. On a visit to the United States in 1929, for example, Werner Heisenberg found

American physicists to be much more receptive to the novelties of the new physics than were his European colleagues. Einstein was similarly impressed, in 1921, with the enthusiasm shown by younger Americans for acquiring knowledge and doing research. He remarked: "Much is to be expected from American youth: a pipe as yet unsmoked, young and fresh." This story of the American reception of the new currents in physics has been ably told by historians Katherine Sopka, Stanley Goldberg, Lawrence Badash, and Daniel Kevles.[5]

What follows is a study of the shifting intellectual commitments of American physicists during the decades immediately preceding the advent of relativity and quantum theory. A review of the contrasting appraisals of physics given around 1880 by John Stallo and his critics establishes a foothold into the thought of the time. Given this reference point, the study then turns to an analysis of the actual intellectual leanings of representative American physical scientists from about 1870 to 1895. Along with reviewing the range of outlooks of those physicists with mechanical leanings, this survey also focuses on those physical scientists who more openly deviated from conventional beliefs. Finally, the study moves to the period between the years 1895 and 1905 and examines the newer viewpoints especially among the younger physicists attending the St. Louis Congress of Arts and Science. Throughout, to be sure, we will see how acknowledged leaders in science play a large role in this story of late nineteenth-century conceptual change. They will appear as professors engaged in teaching and research, as officers of professional societies, and as heads of scientific organizations. Nevertheless, the contribution of lesser known, yet intellectually active physicists, will also be part of the emerging picture. Throughout this study, we will see how a comparative international perspective provides an essential framework for understanding American orientations. By frequently referring to the ideas of European physicists, particularly British, we will be able to identify a number of transatlantic influences, parallels, and contrasts.

STALLO AND THE STATE

OF PHYSICS

Meeting of the National Academy of Sciences, Smithsonian Institution, early 1870s. Participants include Alfred Mayer (10), Simon Newcomb (22), and Albert Michelson (24); Joseph Henry (1) is presiding. (Smith-

1

STALLO'S CRITIQUE

What were the intellectual preferences and leanings of American physicists in the years around 1880? John B. Stallo, a scientifically astute layman from Ohio, believed he knew. Or more accurately, he believed he could characterize the intellectual orientations of the international physics community, the Americans included. Physicists in the United States, like their colleagues in England, Germany, and France, combined in their science an avowed "mechanical" outlook with an unconscious "metaphysical" bias. In Stallo's opinion, this "mechanical" and "metaphysical" orientation was the source of what he perceived to be the practical failings of late nineteenth-century physics.

Drawing evidence mainly from European and occasionally American scientists, Stallo elaborated his adverse appraisal in his 1882 book, *The Concepts and Theories of Modern Physics.* "Modern physical science aims," he began in his first chapter, "at a mechanical interpretation of all the phenomena of the universe. It seeks to explain these phenomena by reducing them to the elements of mass and motion and exhibiting their diversities and changes as mere differences and variations in the distribution and aggregation of ultimate and invariable bodies or particles in space." In other words, physicists sought to understand all phenomena—light, heat, electricity, magnetism, the behavior of gases, and even gravity—in terms of underlying, unseen particles that obeyed established laws of dynamics. According to Stallo, this "mechanical" or, to be more precise, "atomo-mechanical" proposition found wide and deep support. "With few exceptions," he reported, "scientific men of the present day hold the proposition . . . to be axiomatic, if not in the sense of being self-evident, at least in the sense of being an induction from all past scientific experience. And they deem the validity of the mechanical explanation of the phenomena of nature to be, not only unquestionable, but absolute, exclusive, and final." Such talk of unseen atoms and absolute answers, moreover, led Stallo to reject the prevailing assumption that modern, empirical physical science had escaped "from the cloudy regions of metaphysical speculation"; rather, he detected "the insidious intrusion into the meditations of the man of science of the old metaphysical spirit." To the detriment of their science, most physicists shared a "metaphysical" commitment to an "atomo-mechanical" world view.[1]

Was this appraisal accurate? At least a few of the dozen or so American and British critics who initially reviewed *The Concepts and Theories of Modern Physics* thought so. But other reviewers judged that Stallo's appraisal was questionable. Even if physicists were committed exclusively to an atomo-mechanical outlook—a debatable assertion in itself—they were committed to the outlook only in provisional or hypothetical, and hence, nonmetaphysical terms. An atomo-mechanical representation of a phenomenon, emphasized Stallo's detractors, is merely a "working hypothesis." But Stallo refused to concede. After listening to the protests of his reviewers, he published lengthy rebuttals in which he pressed, even more vigorously, his indictment.

Although this intellectual skirmish between Stallo and his American and British critics is, in itself, a revealing episode in Stallo's career, it is also significant for a broader reason. Specifically, the interchange provides a convenient entrée into the milieu of late nineteenth-century physics: the assertions of Stallo and his supporters and the counter-assertions of his critics provide a set of opposing propositions which we can evaluate within the American historical context. To better secure this foothold and establish a framework for subsequent analysis, therefore, it will be useful to examine more closely Stallo's appraisal, his reviewers' responses, and Stallo's rebuttal.

Neither Stallo's profession in the period around 1880 nor the location of his home would indicate that he was the author of a technically sophisticated analysis of modern physics. A lawyer and one-time judge living in Cincinnati, Ohio, Stallo was far from the world's scientific centers. But a closer look at his prior training and interests makes plausible his later achievement. He was born in Germany in 1823 and attended German schools before immigrating to Cincinnati at the age of sixteen. Apparently, it was in Germany that he acquired a background and taste for philosophy and the

[4]

intellectual life. At eighteen, he began teaching German at a new Jesuit college in Cincinnati. During his three years there, he also taught mathematics and took the opportunity to study physics and chemistry. Quickly mastering these fields, he left Ohio for New York, becoming, in 1844, a professor of natural philosophy, chemistry, and mathematics at St. John's College (later Fordham University). While at St. John's, he intensified his reading in philosophy, and in 1848, published a book on Hegel and other German thinkers, titled *General Principles of the Philosophy of Nature*. But rather abruptly he turned to the study of law, eventually interrupting his academic career in 1848 to return to Cincinnati as a lawyer. In discussing this shift, he later recalled: "I desired primarily to make sure of a secure living for the future, so I came back to Cincinnati. I wanted to become practical, as the Americans are." Becoming "practical" meant becoming successful and respected in Ohio as a lawyer, public servant, and for a few years, a judge.[2]

During the three decades between passing the bar examination in 1849 and publishing *The Concepts and Theories of Modern Physics* in 1882, Stallo continued to write essays on a wide range of civic, historical, and scholarly topics, including an 1855 philosophical argument against materialism.[3] He also returned briefly to academic life, teaching chemistry at the Eclectic Medical Institute of Cincinnati. More important, he worked out the rudiments of his later, book-length critique of "mechanical" and "metaphysical" science. These nascent thoughts were presented during 1873–74 in a series of four articles collectively titled "The Primary Concepts of Modern Physical Science" and published in Edward L. Youmans' *Popular Science Monthly*. Ten years later, large sections of these articles would reappear verbatim in Stallo's book.

Stallo apparently was motivated to write these articles as much by a desire to debate prominent analysts of science like John Stuart Mill and Herbert Spencer as by a desire to criticize scientists themselves. This German-born Cincinnatian—whose own empiricist stance against metaphysical thought was distinguished and tempered by Hegelian and Kantian roots—was boldly entering the popular but erudite dialogue carried on by Mill in his *A System of Logic* and *An Examination of Sir William Hamilton's Philosophy* and by Spencer in his *Principles of Psychology* and *First Principles*. Stallo's interest in this type of sweeping scientific discussion prob-

[5]

ably endeared him to editor Youmans, one of Spencer's leading disciples and publicists.[4]

Stallo was also seeking to confront prominent scientists such as British physicist John Tyndall and German physiologist Emil Du Bois-Reymond. Even though Tyndall was at the peak of popularity in the United States following his heralded lecture tour of the nation earlier in 1873 (see chapter 7), Stallo repeatedly chastised him for naively urging "the theory of the atomic constitution of matter as the only theory consistent with its [matter's] objective reality." In Stallo's view, however, the atomo-mechanical outlook found "its most emphatic expression" in the recent writings of Du Bois-Reymond, a leader of the mid-century trend to reduce all analyses of life processes to the principles of physics and chemistry. Even though Du Bois-Reymond had admitted there were human limits to extending the mechanical program, he had equated these particular limits with "the irremovable bounds of all possible knowledge respecting physical phenomena . . . ." Looking at Du Bois-Reymond's dramatic and often-quoted exclamation, *"Ignoramus—Ignorabimus,"* Stallo detected "a dogmatism of ignorance" that was "as presumptuous as the pretended omniscience of a giddy metaphysician." Perhaps men "do not know—will never know," but it was "audacious" to view this limitation solely from the myopic mechanical perspective.[5] Apparently, certain Americans appreciated Stallo's series of articles. Not only Youmans, editor of *Popular Science Monthly,* but also William T. Harris, editor of the *Journal of Speculative Philosophy,* highly recommended the series to their readers.[6]

Building on his articles of 1873–74, Stallo went on to formulate an even more forceful critique of physics—a refined and expanded critique which he presented in 1882 as *The Concepts and Theories of Modern Physics.* He set himself two related goals, both arising from his general perception that physicists were metaphysically committed to an atomo-mechanical view of nature. First, by examining the detailed workings of mechanical physics, he wished to show that this branch of contemporary science was neither internally consistent nor capable of explaining certain basic empirical facts. That is, he wished to show that modern physics was both a logical and practical failure. His second goal was to account for these failings by exposing the shaky, metaphysical foundations of

[6]

modern physics. The latter was, in fact, his main goal. He was using physics primarily as a case study to demonstrate the "logical and psychological origin" of metaphysical modes of thought and cognition—metaphysical modes that were intellectually emasculating but pervasive in nearly all fields of human inquiry. In the first sentence of the Preface, he wrote, "The following pages are designed as a contribution, not to physics, nor, certainly, to metaphysics, but to the theory of cognition."[7]

In line with his first and lesser objective, Stallo devoted the opening eight chapters of his book to a technical examination of whether the atomo-mechanical theory "is consistent with itself and with the facts for the explanation of which is it propounded." On the basis of gleanings from recent scientific writings, he began by describing three assumptions that "lie at the base of the whole mechanical theory" and "command universal assent among physicists of the present day." Physicists believed that, first, "the primary elements of all natural phenomena—the ultimates of scientific analysis—are mass and motion"; second, "mass and motion are disparate"; and third, "both mass and motion are constant." He also found "generally prevalent among physicists and chemists" a belief in "the atomic constitution of bodies." Coupling this atomic assumption with the preceding mechanical ones logically led, in Stallo's opinion, to four additional propositions that "constitute the foundations of the atomo-mechanical theory." These atomo-mechanical propositions—which respectively stipulated that all atoms are identical, inelastic, and inert and that all energy is kinetic—constituted the framework around which Stallo proposed to mold his detailed critique of modern physics. He was planning to isolate, on the one hand, currently functioning hypotheses that contradicted the four basic theoretical propositions and, on the other hand, functioning hypotheses that were consistent with the propositions but not with empirical fact.[8]

The first atomo-mechanical proposition was that the "elementary units of mass . . . are in all respects equal."[9] Whereas logically consistent thinkers—including, in Stallo's opinion, Herbert Spencer—endorsed this proposition, Stallo observed that "physicists, and especially chemists, of our day evince a disposition to ignore this essential feature of the mechanical theory." In particular, Avogadro's law—the chemists' indispensable dictum that

THE CONCEPTS AND THEORIES

OF

MODERN PHYSICS

CHAPTER I.

INTRODUCTORY.

MODERN physical science aims at a mechanical interpretation of all the phenomena of the universe. It
seeks to explain these phenomena by reducing them to
the elements of mass and motion and exhibiting their
diversities and changes as mere differences and variations in the distribution and aggregation of ultimate
and invariable bodies or particles in space. Naturally
the supremacy of mechanics became conspicuous first
in the domains of those sciences which deal with the
visible motions of palpable masses—in astronomy and
molar physics; but its recognition is now all but universal in all the physical sciences, including, not only
molecular physics and chemistry, but also such departments of scientific inquiry as are conversant about the
phenomena of organic life.

It is said that the theoretical no less than the practical progress of the natural sciences, during the last

B

First page of the 1882 edition of Stallo's book. (Permission of Routledge
and Kegan Paul Ltd.)

equal volumes of all gases contain equal numbers of molecules—was "directly subversive of" and in "utter and irreconcilable conflict with" the premise that all chemical atoms are of the same weight. Even if one granted that chemical atoms had different weights, as implied by Avogadro's law, and instead sought uniformity of weight in some smaller, component "atom" or particle, such as William Prout sought in elementary hydrogen, one became trapped in an even greater physical contradiction. According to the physicists' view that heat was a mode of molecular or atomic motion, one would expect, in theory, that as the number of particles within each molecule of a gas increased, there would be a corresponding increase in the gas's specific heat—the heat required to raise the temperature of one unit mass of a substance by one degree. But Rudolf Clausius, Ludwig Boltzmann, and James Clerk Maxwell—three of the pioneers of atomo-mechanical physics—had found that for a gas like chlorine the "calculated specific heat would far exceed the amount ascertained by actual experiment" if one assumed that a "molecule consisted of a number of atoms so great as to be sufficient to account for the differences between the molecular weights of the elements." Thus, on the one hand, Avogadro's law for molecules in a gas was logically inconsistent with the fundamental proposition requiring equality of atoms. But, on the other hand, if one attempted to eliminate this inconsistency by postulating submolecular particles, one moved into conflict with the predictions of the mechanical theory of specific heats. Stallo deplored this conceptual discord.

The actual, functioning research program of physical scientists was also irreconcilable with what Stallo perceived to be the second fundamental proposition of the atomo-mechanical outlook—the proposition that "the elementary units of mass are absolutely hard and inelastic."[10] In particular, the kinetic theory of gases, the "most conspicuous" of modern mechanical hypotheses, described a gas as a conglomeration of moving, colliding, *elastic* particles. Stallo smugly pointed out that "the necessity of attributing perfect elasticity to the elementary molecules or atoms in view of the kinetic theory of gases" had been acknowledged by the theory's founders, including August Kroenig, Clausius, and Maxwell. Similarly, he quoted Sir William Thomson's rejection of inelasticity—and hence of the second proposition—on the grounds of conservation of energy.

Stallo went on to discuss the efforts of leading physicists to avoid this dilemma through "sub-atomic" explanations of atomic elasticity. Most prominent was the popular "vortex-ring" explanation that William Thomson had formulated in the late 1860s on the basis of prior research by Hermann von Helmholtz. Thomson considered an atom to be analogous to a smoke ring. Specifically, an atom was nothing more than a permanent rotational or vortex motion in a perfectly homogeneous, incompressible, and frictionless fluid or ether. Although Stallo admitted that such vortex-ring atoms would indeed interact like perfectly elastic bodies, he presented, nevertheless, two damning criticisms. First, speaking as an empirical philosopher, he argued that "motion in a perfectly homogeneous, incompressible and therefore continuous fluid" is neither "sensible," "phenomenal," nor "real" motion but rather "altogether illusory" and "purely ideal." Second, citing a prior finding of Maxwell, he asserted that vortex-ring atoms could not account for the inertia of matter; it was logically "impossible" to construct both inertia and energy, mass and motion, from ethereal motions alone. In Stallo's opinion, defenders of the vortex theory had failed to propose an atom that was a successful alternative to the elastic atom of the kinetic theory. More generally, defenders of the kinetic theory had failed to resolve the theory's conflict with the proposition of hard, inelastic elementary particles.

Contemporary physicists also contradicted a third basic proposition—that "the elementary units of mass are absolutely inert and therefore purely passive."[11] Although they required, in theory, that all physical action be reducible to the contact or impact of passive masses, physicists had to resign themselves, in practice, to explanations of gravitation that involved action at a distance. Scientists had failed to devise a mechanical explanation of Newton's universal gravitation. Stallo pointed out, nevertheless, that "the old prejudice" against action at a distance as vitalistic or occult "is almost, if not quite, as prevalent now as it was two centuries ago." This prejudice he found in the broad doctrinal assertions of Du Bois-Reymond as well as in those of Balfour Stewart and P.G. Tait, the British coauthors of *The Unseen Universe,* a widely read, speculative essay on the ties between physics and the spiritual realm. The prejudice was most evident, however, "in the incessant renewal, by distinguished men since Newton's day, of the attempts to account for the phenomena of gravitation on the principles of fluid pressure or solid impact." Neither fluid theories (such as that of James Challis involving a continuous, elastic ether) nor impact theories (such as that of G.L. Le Sage involving multidirectional streams of "ultramundane corpuscles") were successful, however, in explaining gravity. Even Stewart and Tait conceded the failure of fluid theories, while Maxwell demonstrated that Le Sage's impact theory violated conservation of energy. Consequently, though the notion of action at a distance was irreconcilable with the third basic atomo-mechanical proposition, the notion persisted in modern physics as, in Stallo's word, "an ultimate fact."

The fourth and final proposition was that "all potential energy, so called, is in reality kinetic." Advocated by Stewart, and particularly by Tait, this proposition was actually a corollary of the previous one.[12] Just as mechanical physicists denied action at a distance with its occult connotations, they also rejected the idea of stored energy of position or potential energy. But, in Stallo's view, even though strict mechanists such as Stewart and Tait voiced the logical necessity of this proposition, most practicing physicists continued to analyze physical phenomena in terms of conversions of kinetic *and* potential energies. Moreover, according to Stallo, the fertility of modern physics actually resulted from the "progressive abandonment" of the proposition that all energy is kinetic, a misleading proposition traceable to Descartes' theory of the conservation of motion. For Stallo, the history surrounding the doctrine of the conservation of energy demonstrated that modern physics had advanced by rejecting this final proposition.

Having spent four chapters of his book discussing the four propositions of *atomo-mechanical* physics, Stallo narrowed his analysis in the next chapter to focus solely on the *atomic* doctrine of modern science. If his return to this fundamental topic seems both a stylistic and logical deviation from what preceded, it is because he inserted this critique into his book almost verbatim from his first article in *Popular Science Monthly* (1873). In this previously published chapter, he identified and analyzed three assumptions held by defenders of the atomic doctrine, including primarily the assumption that "matter consists of discrete parts, the constit-

[11]

uent atoms being separated by void interstitial spaces."[13] This assumption of the discontinuity of matter was essential to advocates of the undulatory or wave theory of light as formulated in the early 1800s by Thomas Young and Augustin Fresnel. The dispersion of light, for example, was inexplicable in terms of a luminiferous or light-propagating ether that not only was elastic (to account for the transverse oscillations of light) but also continuous; as Augustin Cauchy had shown earlier in the century, one needs to assume that "the aethereal medium of propagation, instead of being continuous, consists of particles separated by sensible distances." Such a granular ether accounted for the spectrum of colors in dispersion since the velocities of propagation of the different colors would vary according to wavelength.

Stallo disagreed. He believed there were two reasons for questioning if dispersion could be satisfactorily explained by this granular ether. First, as the American engineer E.B. Hunt emphasized in a article published in 1849 in Silliman's *American Journal of Science,* various astronomical phenomena revealed that component colors of the light spectrum were not in fact propagated with unequal velocities. Cauchy's theory of variable velocities, therefore, was simply "fallacious," "at variance with observed fact." Second, as Maxwell argued—and this was a point Stallo added since writing his original article of 1873—a granular ether would behave essentially like a gas and therefore display "a correspondingly enormous specific heat . . . ." Since ether permeated vacuums, these regions of otherwise empty space would have a specific heat. This was a "remarkable consequence," in Stallo's view, "without experimental warrant." A similar difficulty ensued if this granular ether filled the spaces between the molecules of a gas; one would calculate, according to the kinetic theory of gases, a value for the specific heat of the gas even higher than the regular theoretical value—a regular value that was already in trouble since it exceeded what had been determined by experiment. In Stallo's view, he had discredited Cauchy's attempt using a granular ether to include dispersion in the undulatory theory of light; consequently, advocates of the atomic doctrine could not point to dispersion in justifying the assumption of the discontinuity of matter.

In concluding the first half of his book, Stallo returned to a general discussion of the kinetic theory of gases. To set the stage for

[12]

this closing critique, he first inquired into the criteria for a useful and valid scientific hypothesis.[14] "A scientific hypothesis," according to Stallo, "may be defined in general terms as a provisional or tentative explanation of physical phenomena." The adjectives "provisional or tentative" were important for Stallo. Earlier, for example, he had emphasized that his criticisms of the atomic theory "do not, of course, detract from the merits of the atomic hypothesis as a graphic or expository device . . . ."[15] But to be valid, an hypothesis had to satisfy two conditions: "the first of which is that every valid hypothesis must be an identification of two terms—the fact to be explained and a fact by which it is explained; and the second that the latter fact must be known to experience." Stallo qualified, however, adding that "the identification of the two phenomena may be both partial and indirect." That is, as long as a scientist maintained some known ties between the two phenomena, he could assume the existence of "some other feature not yet directly observed, and perhaps incapable of direct observation." Nevertheless, "the probability of the truth of an hypothesis" decreased with the proliferation of such "fictitious elements." In the extreme case, where there were connections only between elements that were purely fictitious, the hypothesis became "wholly vain" and "meaningless—a mere collection of words or symbols without comprehensive import." Such a "multitude of fictitious assumptions" occurred, for example, with the undulatory theory of light and its luminiferous ether. Stallo reported "an impulsive whirl in our thoughts" when asked to accept an all-pervasive luminiferous ether that exerted a tremendous elastic force yet was both non-resistent to ordinary moving bodies and completely unobservable. He also reported being dumbfounded when asked to acknowledge other distinct ethers, such as the "electriferous" one, which would account for phenomena inexplicable with the already complex luminiferous ether. Incidentally, Stallo had been even more critical of ether in his articles of 1873–74, characterizing it as ". . . nothing but a clothes-horse for all the insoluble difficulties presented by the phenomena of sensible material existence—a fagot of occult qualities . . . ."[16]

Drawing on his criteria for hypotheses given in this methodological digression, Stallo quickly concluded that "the kinetic hypothesis has none of the characteristics of a legitimate physical

[13]

theory."[17] Returning to the issue of atomic elasticity, for example, he ridiculed attempts to account for the relatively simple phenomenon of the elasticity of a gas by postulating extremely complex elastic atoms: "It is utterly vain; or rather, inasmuch as it complicates the phenomenon which it professes to explicate, it is worse than vain—a complete inversion of the order of intelligence, a resolution of identity into difference, a dispersion of the One into the Many, an unraveling of the Simple into the Complex, an interpretation of the Known in terms of the Unknown, an elucidation of the Evident by the Mysterious, a reduction of an ostensible and real fact to a baseless and shadowy phantom."

Stallo also pointed to specific difficulties with the kinetic theory. He found Maxwell's early assumption, that gas molecules repel each other with a force inversely proportional to the fifth power of their distance, to be "purely gratuitous," "without experimental analogy," and "in direct defiance . . . of actual observation." Also dissatisfied with the "methods of deduction" used in kinetic theory, he further chided Maxwell for resorting to "the method of statistics" in calculating gas pressures and other effects. Noticing Maxwell's emphasis in an 1874 *Popular Science Monthly* article on the "velocity of mean square" of molecules, Stallo rhetorically asked, "Now, on what logical, mathematical, or other grounds is the statistical method applied to the velocities of the molecules in preference to their weights and volumes?" Finally, Stallo confronted "the most important witness yet called on behalf of the kinetic theory," the spectroscope. The complexity of spectra of gases seemed to suggest that molecules have a correspondingly complex internal structure of vibrating particles; moreover, the persistence of the characteristic spectra seemed to suggest an absence of any "clashing of the molecules" that might disturb "the natural periods of vibration." In opposing this alleged evidence for a substratum of kinetic particles, Stallo enlisted, for a change, the support of Maxwell. If, as the spectroscope implied, the molecules of a particular gas were internally complex, then, as Maxwell realized in 1875, the specific heat of the gas should be higher than was normally calculated for a molecule consisting of only a few simple atoms. But, as previously mentioned, this calculated specific heat was already greater than the measured value. In other words, observed specific heats could not be reconciled with a simple molecular structure, let alone with the complex internal structure implied in spectral evidence. For Stallo, problems such as these invalidated the kinetic hypothesis. Taking one last overview of this theory of gases, he drew a parallel between the study of kinetic-molecular phenomena and spiritual beings: "the intangible specter proves more troublesome in the end than the tangible presence. Faith in spooks (with due respect be it said for Maxwell's thermo-dynamical 'demons' and for the population of the 'Unseen Universe') is unwisdom in physics no less than in pneumatology."

In the second half of *Concepts and Theories of Modern Physics,* Stallo sought to account for the failings of the atomo-mechanical program by exposing its shaky, metaphysical base. Through an inquiry into the "logical and psychological origin" of the program, he sought to discredit "the distinct claim of modern physicists that the mechanical theory rests on the sure foundation of sensible experience, and is thus contradistinguished from metaphysical speculation . . . ." He defined "metaphysical thinking" as "an attempt to deduce the true nature of things from our concepts of them." Such fallacious thinking usually took the specific form of one of four naturally occurring "structural fallacies of the intellect." For instance, the first structural fallacy, "that every concept is the counterpart of a distinct objective reality," was evident in how physicists treated the concepts, "matter," "motion," and "mass." After reviewing the customary identification of these concepts "with real, sensible objects," Stallo concluded that "the mechanical theory . . . hypostasizes partial, ideal, and it may be, purely conventional groups of attributes, or single attributes, and treats them as varieties of objective reality. Its basis, therefore, is essentially metaphysical. The mechanical theory is, in fact, a survival of medieval realism." Similarly, he detected in modern physics a fourth "structural fallacy" or "radical error of metaphysics"—"that all physical reality is in its last elements absolute." This particular "ontological prejudice" constituted, in Stallo's opinion, "the true logical basis" of the atomo-mechanical theory. The prejudice permeated modern physics in such guises as absolute space, time, motion, and rest as well as absolute material substance, physical unit, and physical constant.[18]

As he had done earlier in his articles of 1873–74, Stallo continued throughout this last half of his book to dissect the prior arguments of prominent analysts of science, particularly Mill, and to extract damning evidence from the writings of contemporary scientists, particularly Tyndall. Such an intensive examination of the philosophical and psychological roots of modern physics, coupled with the earlier technical analysis of its logical and practical failings, led Stallo finally to conclude:

> the atomo-mechanical theory is not, and can not be, the true basis of modern physics. On proper examination, this theory . . . proves to be . . . incompetent to serve as an explanation of the most ordinary cases of inorganic physical action. And the claim that, in contradistinction to metaphysical theories, it resorts to no assumptions, and operates with no elements save the data of sensible experience, is found to be wholly inadmissible.[19]

Stallo, lawyer at heart, had completed his initial prosecution; it was now time to hear from the counsel for the defense.

2

REACTIONS OF REVIEWERS

Soon after the appearance of *The Concepts and Theories of Modern Physics,* more than a dozen American and British periodicals carried reviews of the book. It commanded attention in part because of its iconoclastic tone. Harvard philosopher Josiah Royce later recalled that with the publication of Stallo's book the "sense of scientific orthodoxy was shocked amongst many of our American readers and teachers of science."[1] Moreover, the book commanded attention because it was published in the United States and in England as a volume in the "International Scientific Series." This was a prestigious series initiated by Edward Youmans and published by the Appletons; Stallo, the thirty-eighth author, was preceded by such prominent scientists as John Tyndall, Balfour Stewart, Josiah P. Cooke, and J. Norman Lockyer.[2]

Depending on the profession of each individual reviewer, the responses ranged from praise to scorn. In most cases, if the reviewer was a practicing research scientist, he tended to reject Stallo's arguments. On the other hand—although certain supporters of Stallo's book were anonymous—if a reviewer seemed to have been a popular essayist, humanist, armchair scientist, or even an engineer writing for a lay audience, he tended to endorse Stallo's critique. For example, the unnamed reviewer in *The American,* a "national weekly journal of politics, literature, science, art, and finance" published in Philadelphia, pronounced Stallo's book to be a "timely-happy" response to the "shortcomings" of modern science. Similarly, an unnamed writer for *The Literary World,* a "fortnightly review" originating in Boston, felt that "no volume more interesting and instructive has yet appeared in the 'International Scientific Series.'" The unidentified reviewer for the *American Engineer,* a Chicago weekly founded in 1880, was even more enthusiastic. "It is not exaggerating the importance of this work," he began, "to say that it ranks among the remarkable contributions to scientific thought of this century and will have a

[17]

marked influence on the attitude of scientific inquiry towards its object."[3]

Stallo found his strongest backing, however, in the pages of *Popular Science Monthly,* the journal in which he had presented his earlier articles. It was published by the same two men who had produced his book, editor Youmans and publisher Appleton. Four times during 1882, ex-chemist Edward Youmans lent the support of his widely read, popular-level journal to the new book. Youmans began by calling "especial attention" to the upcoming publication of this book of "exceptional importance." In previewing Stallo's discussion of atomo-mechanical physics and "the line between legitimate science and illegitimate speculation," the editor concluded "there is now no subject that more urgently needs to be cleared up than this." A month later, he ran a full book review that praised Stallo for discussing "accepted theories with great critical skill and logical force, showing their short-comings and contradictions, and proving conclusively that what is now most needed is a thoroughgoing re-examination of the grounds of what is currently regarded as established scientific theory."[4] He then reprinted a review that had originally appeared in the *Canadian Monthly* endorsing Stallo's more philosophical points. Finally, Youmans provided Stallo with the first twenty pages of the June issue in which to respond to negative reviews that had appeared in other journals, primarily to one written by Simon Newcomb.[5]

Of the fourteen appraisals of *Concepts and Theories of Modern Physics* dating from 1882 and early 1883, six were endorsements, echoing and elaborating Stallo's major points. Of the remaining eight appraisals, five were openly antagonistic, two vacillated between support and rejection, and one was opposed for only nonscientific, philosophical reasons.[6] Practicing scientists wrote each of the five antagonistic as well as the two mixed reviews. Negative reviews came from the pens of British physicists Donald MacAlister and P. G. Tait, American physical scientist Simon Newcomb, and two anonymous American authors—one positively identifiable as psychologist G. Stanley Hall and the other very likely physicist Alfred M. Mayer.[7] The mixed reviews were written by English physicist Arnold W. Reinold and American medical-chemist Robert G. Eccles.[8]

These reviewers poured forth a wide range of objections to all

[18]

phases of Stallo's critique. There were, however, basically four kinds of response. First, they argued that earlier thinkers had anticipated Stallo's critique of both the technical failings and the metaphysical foundations of modern physics. For example, the reviewer for the New York based *Critic*—probably Alfred Mayer, well-known physicist at the Stevens Institute of Technology in New Jersey—maintained that Stallo had set forth no difficulty with either the undulatory theory of light or the ether "which has not been much better displayed in the pages of scientific memoirs." Similarly, Donald MacAlister—senior wrangler at Cambridge in 1877, one of Maxwell's students, and later a physiologist—wrote in the London quarterly *Mind* that Stallo's criticism of the chemical atomic theory and the kinetic theory of gases "are often shrewd even if they are not new." Of course, neither Mayer nor MacAlister were suggesting that scientists had anticipated all aspects of Stallo's technical critique. There were portions (most notably, his attack on Maxwell's use of statistics in kinetic theory) that the reviewers simply dismissed as wrong.[9]

As for Stallo's philosophical analysis of the metaphysical foundations of modern physics, certain reviewers also believed that this was an old and exhausted perspective. Granville Stanley Hall, a scientifically and philosophically astute psychologist who had recently obtained his doctorate from Harvard and was just beginning his career at Johns Hopkins, adopted this mode of criticism in his unsigned review for *The Nation*. He felt that Stallo was adding "nothing whatever" to the modern "theory of cognition" if he were merely saying, "Science is at bottom metaphysics, with a persistent tendency to 'reify' its concepts, which must then be extirpated as ontological survivals." Hall concluded that Stallo "only retails the residual but tedious truism of idealism, that things are really thoughts. . . ." and he derided him for "feeling complacency in reviving a method applied repeatedly a decade or two ago in this field."[10]

In speaking of this method used in previous decades, Hall was probably referring to the "positive philosophy" of Auguste Comte. Earlier in fact, Hall had called attention to Stallo's lack of "allusion to Comte even where it would admirably serve his purpose." Another of Stallo's reviewers—one of his supporters—also believed that Stallo was often merely restating the arguments of

Comte and, occasionally, those of Herbert Spencer. This supporter of Stallo juxtaposed quotations from Comte and Stallo to demonstrate the similarity of their arguments against attempts to reduce gravitation to the impact of passive elementary particles. Having summarized Stallo's version of the argument, the reviewer concluded: "This whole passage is so completely on the lines of the Positive Philosophy, that to us it seems singular that the author could have penned it without making some reference to the precisely similar views of Auguste Comte, views which the scientific world in general has largely disregarded or ignored."[11] When Stallo responded to this particular complaint in the lengthy introduction to the second edition of his book in 1884, he contended that he was indebted to Comte only indirectly. He was obliged only to the extent that Comte's writings "are simply part of the modern theory of Cognition, whose fundamental principles are so familiar to those who are acquainted with that theory that no one, at this day, thinks of citing authority for them." He added that he had "not looked into any of Comte's writings for more than twenty years." Due to forgetfulness or pride, Stallo did not mention that ten years earlier, in his *Popular Science Monthly* articles, he had cited a particular section of Comte's *Philosophie Positive* to bolster his attack against reducing gravitation to the impact of passive particles.[12] As will become apparent, however, this failure to credit a source was somewhat out of character for Stallo.

The second general objection voiced by Stallo's reviewers was that his attack was aimed at strawmen or, to be more exact, straw-physicists. The physicists guilty of the conceptual sins alleged by Stallo, the reviewers protested, were either few in number, unimportant persons, or outright deviants. Drawing the strongest reaction was Stallo's assertion that he had identified four propositions constituting the foundation of modern atomo-mechanical physics. Pointing to the meager and questionable list of scientists Stallo had marshalled as supporters of the proposition regarding the equality of atoms, MacAlister insisted: "We must in fairness ask for better proof that the doctrine is widely held or held at all before we own that our withers are wrung." Similarly, MacAlister reported that modern physics rejects the proposition that atoms are inelastic. In actuality and as Stallo also admitted, "two of the most conspicuous hypotheses of the science—the Kinetic theory of gases, and the

[20]

Vortex-theory of the atom—take account of the elasticity of the ultimate particles, and go far to explain it." Mayer agreed that a few physicists undoubtedly felt that all potential energy was in reality kinetic, but this did not make the belief a fundamental proposition. "It is true that some distinguished names may be cited in support of the former statement," Mayer wrote, "but the same may be said of every fallacy and every heresy. Surely this establishes no paradox."[13]

Hall likewise disapproved of Stallo's "vicious dialectic habit of . . . bumping friendly heads together." Accordingly, Hall stressed that many ad hoc or ancillary hypotheses "which investigators have found useful to meet the exigencies of research—enchainments, thermalizations, valencies, and various dynamisms, etc.—are confessedly not unanimously agreed upon, and there is much residual obscurity which is, of course, increasing as the field of science widens." MacAlister summed up the reviewers' dissatisfaction with Stallo's strawmen when he presented "the answer that any physicist who is familiar with the best and latest teachings in his subject is bound to make to many of Mr. Stallo's criticisms." The answer was simply that "the chief points made against the Atomic Theory refer to unessential or provisional or discarded parts of it. Its best expounders would yield up most of them without feeling that the central doctrines were at all weakened."[14]

MacAlister's mention of the provisional components of the atomic theory leads to another of the objections frequently voiced by Stallo's reviewers. Physical scientists could not possibly have a metaphysical or ontological commitment to the atomo-mechanical theory, since they viewed the theory in a provisional or hypothetical light. Hall, drawing on his experiences with leading scientists at Harvard and Johns Hopkins as well as in Germany, presented the most thorough statement of this objection. For most physicists and especially chemists,

atomism is a didactic, formal, explanatory device, so serviceable that . . . a confessedly wrong auxilliary hypothesis often continues to be of great use. The chief unanimity concerning it is not in the form our author calls axiomatic, but as a symbolic and later a graphic system which might be almost visualized illustratively for specific purposes, but which should not be too grossly conceived, and from which deductions should be very sparingly made for fear of catastrophes. . . . [Atomism] is a complex grammar of facts, or rather a language in which thought instinctively expresses and grasps them, and, like language, its very inconsistencies are often the most valuable and suggestive.

In Hall's opinion, Stallo had not understood the efforts of many leading scientists "to keep a 'fine lucidity' of hypothesis from lapsing into either empiricism on the one hand or dogmatism on the other." Nor had Stallo "really comprehended the newer problems and understood the laboratory function of an hypothesis, both of which tend so strongly to make the theoretic part of science more and more flexible, transparent, formal, and practical. . . ." Newcomb, in his unequivocal rejection of Stallo's book, and Eccles and Reinold, in their mixed reviews, adopted similar stances. After acknowledging Stallo's objections against the kinetic theory of gases and the undulatory theory of light, physicist Reinold, for example, still concluded that such theories "may not be true; but they have proved themselves to be admirable working hypotheses, and in this lies their value."[15]

The fourth and final objection correlates closely with the prior objections that Stallo used hackneyed arguments, attacked strawmen, and misunderstood the provisional nature of physics. Several reviewers concluded that because Stallo himself was not a practicing scientist, he was unqualified to evaluate and criticize modern physics. MacAlister, the main British elaborator of this objection, granted that Stallo was exceptionally well-versed in the literature of modern physics but, as a nonscientist, could never fully understand that literature. There were subtle but essential differences between what scientists said and what they did:

If people always used words accurately and with the same meanings it might be safe to take the words as corresponding with things. But it is dangerous to deal in words exclusively, even when they are the words of approved men of science. Those who have dealt in the things that are faintly shadowed by the words have the clue to their meaning, and from within can read into them much that they fail to express. Hence working men of science understand each other's language, even when it is imperfect, a thousand-fold more clearly than the clearest-headed of those who are hearers only. Mr. Stallo has been a diligent and painstaking hearer and reader. We are convinced he would never have written at least the first half of his book if even to a small extent he had also been a doer. His criticisms apply not to the concepts of physics in the truest sense, but, where they are good, to the verbal expression of them.

To illustrate this point, MacAlister alleged that Stallo neither had a deep knowledge of mathematics nor had ever witnessed the action of vortex-rings formed by smoke; thus, he could not appreciate that the vortex-theory of the atom was a *"real* theory."[16] MacAlister also felt that Stallo, as a nonscientist, displayed "the further fault . . . of not distinguishing between first-rate authorities, whose dicta are worthy of the carefullest consideration, and third-rate or fourth-rate writers, whose opinions can scarcely be called science." He added, parenthetically, that he had noticed this fault "before in American writers." Apparently, to understand modern physics fully, Stallo along with other Americans, needed to know the unwritten politics of the physics community. They needed to know who was "entitled" to say "what is accredited doctrine and what paradox." P. G. Tait echoed MacAlister's general sentiments regarding the misinterpretations of nonscientist Stallo. Tait, one of Britain's leading advocates of purely mechanical physics, worried that a scientifically inexperienced person might unwittingly accept Stallo's superficial arguments. Specifically, "some ardent student, unversed in laboratory work and with no great knowledge of physical principles" might fall "an easy victim" to Stallo's book.[17]

In stating that science must be experienced to be understood, Tait and, particularly, MacAlister were suggesting that a fundamental component of the content of physics was knowable only through personal encounters that could not be verbally articulated. Only "working men of science" could adequately understand each other's language. Seemingly, MacAlister and certain other physicists felt no compulsion to make such knowledge explicit. As Alfred Mayer wrote in response to Stallo's critique: "A sound digestion has little self-consciousness of the operations of the stomach; the sound thinker gives himself but little uneasiness respecting 'the laws of thought.'"[18]

[23]

Even though American reviewers agreed with the general British criticism that layman Stallo was unqualified to evaluate physics, two of these reviewers had a different specific perception of the everyday workings of physics. Hall and Newcomb held that the modern trend in physical science was not toward the private, implicit knowledge that MacAlister described but toward explicit definitions of fundamental concepts in terms of concrete measurements— what Percy W. Bridgman in the 1920s would label "operational definitions."[19] Consequently, rather than criticizing Stallo for being unaware of tacit, unspoken meanings of concepts, Hall and Newcomb criticized him for avoiding what later generations of physicists would call fully articulated operational definitions.

That is, on the broadest level, Hall echoed MacAlister's general objection by calling attention to layman Stallo's "insuperable difficulty . . . in fully grasping the fundamental ideas of science by solitary reading"; Stallo's book was suitable only for "non-experts who deliberately prefer comprehensiveness to depth and accuracy of view." Like MacAlister, Hall also objected to Stallo's "absence of discrimination among authorities, Helmholtz and Maxwell being in no wise distinguished for his readers from the fanatical Dühring and Zöllner, or from the dreamy Spiller." Similarly, on a general level, Newcomb warned Stallo's readers that "the concepts of science are matters of slow growth in the mind, which can reach their fully developed maturity only by a regular course of special training and in a peculiarly genial soil." On a more specific level, however, British and American reviewers diverged. Newcomb went on to charge that Stallo's "total misconception of the ideas and methods of modern science" was evident in his avoidance of clear, operational definitions. Stallo's use of words such as mass, motion, and matter, for example, has "no definite meaning" because he was seemingly unaware of a basic precept: "The very first necessity of any exact scientific proposition is a definition, without ambiguity, of a precise method in which every quantitive measure brought in shall be understood. The conclusions are then valid, assuming that particular method of measurement, but they are not valid on any other method." Hall also concluded that Stallo's "employment of the word *mass* to mean weight, inertia, or even matter, when it has no sense till one has been assigned it for a specific purpose, may perhaps be taken as a type of a general

[24]

error. . . ."[20] Thus whereas MacAlister and Tait condemned Stallo for being a layman out of touch with the implicit, tacit knowledge of scientists, Newcomb and Hall condemned him for being a layman out of touch with the trend toward explicit, operational definitions. Stallo's critics, however, were not to have the last word on any of these points.

3

THE LAWYER'S REBUTTAL

A lawyer accustomed to confronting adversaries in the courtroom, Stallo quickly subjected the claims of his reviewers to "a counter-critical examination." He replied first to Simon Newcomb and, also, with less emphasis, to the anonymous reviewer for the *Critic,* Alfred Mayer. This refutation was published as the lead article in the June 1882 issue of *Popular Science Monthly.* In 1884, he presented an even more comprehensive rebuttal in a forty-page introduction to the second edition of his book. This he directed at the entire body of his detractors.

In answer to the first general objection that physical scientists had anticipated the valid aspects of his critique, Stallo disarmed his reviewers by agreeing with them. In fact, except for occasional oversights such as his failure to credit Comte, he had openly acknowledged his forerunners throughout the book. Even in his "Preface," he had emphasized that his ideas were not necessarily new:

> I desire to say that there are suggestions, in many of the utterances of the men of science here referred to, of a growing sense of the questionability of some of the elements of their scientific faith. I have taken frequent occasion, in the progress of my discussion, to point to these suggestions, to the end of showing that my thoughts are, after all, but the inevitable outcome of the tendencies of modern science, and are, therefore, rather *partus temporis quam ingenii* [the fruits of this epoch rather than of my own invention].

In his 1882 rebuttal, he reinforced this disclaimer. He explained, for example, that Maxwell's discussion of the difficulties with the kinetic theory of gases and Harvard chemist Josiah P. Cooke's discussion of the problem with the luminiferous ether both resembled his own critique. On a more general level, Stallo further surmised that his personal contribution had been merely to consolidate a "considerable amount of scattered material ready to the hand" into an "orderly and systematic" analysis of modern physics.[1]

POPULAR SCIENCE MONTHLY.

JUNE, 1882.

ЗPECULATIVE SCIENCE.

BY J. B. STALLO.

" Wenn ein Kopf und ein Buch zusammenstossen, und es klingt hohl, muss es denn immer das Buch gewesen sein ? "—LICHTENBERG, *the Physicist.*

THE above title is prefixed to an article contributed by Professor Simon Newcomb to the April number of the "International Review." The avowed object of that article is to discredit a recent volume of the "International Scientific Series" ("The Concepts and Theories of Modern Physics ") as a publication unworthy of the company in which it appears, and to denounce its author as a person ignorant of the subject whereon he writes—as a scientific, or rather unscientific, "charlatan " and "pretender" belonging to the class of "paradoxers" whom Professor De Morgan has immortalized in his famous "Budget." I am fully aware that, as a rule, it is both unwise and in questionable taste for an author to make direct reply to criticism, however hostile, baseless, or absurd. The merits of a book must find their vindication, at last, in its contents, and the chief function of the critic is to bring them to the attention of the reader, the value and spirit of the critical performance being of secondary importance. But the case in hand appears to me to be an exceptional one. The unmistakable intent of Professor Newcomb's "criticism" (and, if it be left unchallenged, its probable effect) is to signalize the contents of the book with which he deals as mere drivel, and unworthy of a moment's serious attention. And he writes for a magazine, the majority of whose readers, however intelligent they may be, can hardly be expected to possess that familiarity with the matters under discussion which is a necessary prerequisite to the formation of an independent and trustworthy judgment. All they are likely to know and care is,

Stallo's reply to Newcomb, the opening page of the lead article in *Popular Science Monthly*. (Permission of Times Mirror Magazines, Inc.)

To his reviewers' second main complaint—that his attack was aimed at strawmen—Stallo reacted more aggressively. As evidence that there was wide acceptance of the basic atomo-mechanical propositions, he pointed to the many "persons of the highest scientific authority" who were mentioned in his book. He had drawn on the statements of Thomas Graham, Wilhelm Wendt, Hermann von Helmholtz, James Clerk Maxwell, Carl Ludwig, and, most frequently, Du Bois-Reymond and P. G. Tait. Stallo stressed that "the number of citations from the writings of eminent men of science in support of the propositions . . . might have been extended." To the list of those who defended Tait's proposition that all energy is ultimately kinetic, for example, he added the name of George Barker, a physicist at the University of Pennsylvania.[2]

Stallo was adamant about the wide acceptance of the basic mechanical propositions; he was also confident that many physicists contradicted these propositions in the actual practice of their science. In his 1882 rebuttal, he updated his evidence for this inconsistency by citing recent assertions, including those by William Thomson, that atoms are regarded as *elastic* particles in the kinetic theory. But the very prevalence of these types of assertions by men such as Thomson constituted one of the severest challenges for Stallo. MacAlister recognized this difficulty when he asked: if so many prominent physicists in everyday practice deny certain of the so-called basic propositions, what real significance do the propositions have to physicists? Stallo phrased a cautious reply to this query:

Now, Mr. MacAlister will, no doubt, be very much astonished when I tell him that this accords precisely with what I say, viz., that these propositions are *not* held by the great body of scientific men who theorize in the presence of the facts, for the simple reason that the facts are inconsistent with them. What I maintain is, that the majority of physicists hold a general doctrine which I designate as the atomo-mechanical theory, from which these propositions inevitably follow; but that, when they construct their special theories, either by generalizing the facts of experience, or by framing hypotheses to account for them, they are constrained to discard and repudiate that doctrine.[3]

Whereas MacAlister perceived that in actual practice most physicists simply denied or disregarded Stallo's propositions, Stallo perceived that most physicists held outlooks that were fundamentally ambivalent or inconsistent.

[28]

The reviewers' third general criticism was their most penetrating: since physical scientists viewed the atomo-mechanical theory in a provisional or hypothetical light, they could not possibly have a metaphysical or ontological commitment to it. Put more simply, Stallo had overlooked the essentially heuristic role of theories. Facing this accusation, Stallo was dumbfounded. In light of his explicit acknowledgment of the necessity of "working hypotheses," he was "wholly at a loss to see what justification" there was in the charge. He quoted extensively from his book to demonstrate the misrepresentation and apparent oversight of, particularly, critic G. Stanley Hall.[4]

Stallo went on, moreover, to confront the more subtle question of how a scientist could be metaphysically committed to an admittedly provisional theory. Here he replied that even the tentative hypotheses of a scientist could insidiously entail tacit metaphysical biases. Regarding the invention, for instance, of diverse and often contradictory ad hoc hypotheses concerning either the luminiferous ether or the atom, he concluded:

> When this is done with a proper insight into the nature and use of such fictions—with the understanding that they are mere devices for fixing ideas or colligating facts (to use Whewell's expression)—it is well enough. But, in many cases, the specialists have no such insight. They begin to treat the fictions here spoken of as undoubted realities, whose existence no one can question. . . .

Similarly, whereas Stallo accepted the heuristic value of even an admittedly defective working hypothesis, he objected to the scientist who sought "to obtrude his own particular hypothetical figment as a finality upon science generally, and to make it the basis of assertions respecting the ultimate constitution of things, and the universal order of nature." In his 1882 rebuttal to Newcomb, he underscored this objection with an analogy involving money—an analogy chosen perhaps to pique Newcomb, an occasional contributor to economic theory. As long as a person remembers that coins made of base metal have no intrinsic value, he can use the coins for "good purpose as mere counters or tokens"; most persons, however, forget that the coins are worthless as metal and instead "insist dogmatically that the coins are of unquestionable intrinsic merit." From this analogy, Stallo drew a moral for scien-

[29]

tists. "If you are unable to procure genuine theoretical specie to represent the scientific wealth you are intent on accumulating, and at the same time are unwilling to restrain your propensities for manufacturing spurious coin and palming it off on yourself and others as sterling cash, you had better carry your facts about in baskets or bags, and resort to the ancient clumsy method of barter." Although he did not name them, Stallo insisted that there were an increasing number of "thoughtful physicists" in agreement with him. These physicists were willing to discard, for example, the atomic hypothesis and to evince the phenomenological spirit displayed earlier in the century by German physicist Gustav Kirchoff. Stallo detected "a growing tendency to divest the edifice of physical science as far as possible of the hypothetical scaffolding, which not only obstructs the view of its fair proportions, but masks the real principles of its construction, by which its strength and permanence are assured."[5]

Even though Stallo was wary of atomo-mechanical hypotheses with their metaphysical potential, he believed that certain other hypotheses were pointing toward valid statements of scientific truth. For example, he was enthusiastic about the principle of conservation of energy. There was "little doubt" that this principle would prove to be "the great theoretical solvent of chemical as well as of physical phenomena."[6] This general belief that nature was ultimately transparent to proper scientific inquiry undergirded Stallo's rebuttal to the fourth and final of his critics' objections—that Stallo as a layman was unqualified to evaluate science.

Because hypotheses were not merely heuristic, arbitrary inventions, but guesses at "ultimate truth" and candidates for "ultimate validity," and because science thus had profound philosophical import, Stallo felt a responsibility to aid physics with the "modern theory of cognition" and the "sciences of comparative linguistics and psychology." Why should physicists be concerned with what layman Stallo had to say? According to Stallo:

> The simple answer to this is, that physical theories are not merely instrumentalities for the discovery and classification of facts in furtherance of the practical purposes of life, but that they also serve as a basis for the various attempts at a solution of the great questions which always have been and always will be the cardinal problems of human thought. And before these questions can be properly submitted to the arbitrament of the physi-

cist, it is necessary that he should have a clear insight into the nature of the concepts and theories with which he operates, and into their relations to the phenomena which they represent.[7]

Having a mastery of the "theory of cognition," Stallo felt both qualified and obligated to provide this clear insight.

In his rebuttals of 1882 and 1884, Stallo was not content merely to refute the arguments of his critics. Like a good lawyer making a closing statement, he soon seized the offensive and used the critics' very words to vivify points of his own original critique. In particular, he called attention to the atomo-mechanical biases that American physical scientists Newcomb and Mayer revealed, along with Scotsman MacAlister, in their reviews. Thus, he sarcastically commented in his 1882 rebuttal: "It is not a little instructive to note the character of sacredness ascribed by persons of Professor Newcomb's frame of mind to dominant physical theories, and the violence with which they repel every attempt to point out their defects. My reviewer in 'The Critic' is almost beside himself after reading my 'assault' on 'that magnificent fabric of science, the undulatory theory of light and heat.'" Mayer, the probable reviewer in the *Critic,* had also defended the kinetic theory of gas as "the grandest of modern inductions."[8]

Stallo developed this same type of counterattack in his 1884 rebuttal. For "proof" of scientists' belief in the "ultimate validity" of certain theories and concepts, he urged his readers to notice "the violence with which my 'assaults' upon the atomic theory and the kinetic theory of gases . . . have been repelled." In this later rebuttal, he concentrated mainly on MacAlister's biases. MacAlister had left himself vulnerable by vigorously endorsing the basic propositions that atoms are absolutely inert and that all energy is kinetic. Specifically, MacAlister had contended that "physics is theorizing in the line of truth" in its ongoing efforts to explain gravity in terms of the impact of inert corpuscles. Moreover, regarding the proposition that all energy is kinetic, he had remarked that successful physicists "are each and all inspired with the faith that this fundamental mechanical proposition is true." Stallo also derided MacAllister for trusting that the vortex-theory of the atom would eventually account for gravitation and inertia and for insisting that vortex atoms in an homogeneous ether could produce "sensible motion." This allegiance to the vortex-atom theory was, in Stallo's

[31]

view, "characteristic of the confusion of modern theorists, who insist upon reducing all physical action to impact. . . ."[9] Stallo concluded that MacAlister, like his fellow physical scientists Mayer and Newcomb, was irremediably committed to an atomo-mechanical world view.

Stallo and his critics offered a variety of contrary characterizations of late nineteenth-century physics;[10] these divergent claims provide a convenient set of opposing propositions that may be further evaluated within the context of the American physics community. We can summarize these main assertions and counter-assertions using four overlapping groups of questions that build on the four basic responses of Stallo's critics. First, were American physical scientists aware of the technical and metaphysical failings of the atomo-mechanical theory—failings that Stallo detailed, and that some of his critics readily acknowledged? In other words, were portions of Stallo's critique passé by the 1880s? Second, did American physicists hold fundamental propositions similar to those Stallo had listed? If there were such men—a circumstance which Stallo's critics doubted—did they endorse the propositions only in theory, repudiating them, as Stallo maintained, in actual practice? That is, were American physicists ambivalent and logically inconsistent in their mechanical outlook?

Third, were American physical scientists metaphysically committed to the atomo-mechanical theory? Or were these scientists, as Stallo's critics responded, committed to the theory only provisionally or tentatively? And even if their outlook was avowedly provisional, did the outlook nevertheless entail, as Stallo believed, a subtle ontological bias? Finally, were Americans failing to tie their concepts and theories adequately to the observational realm, or was this merely the misperception of a layman out of touch with actual scientific practice? Was there in fact a trend toward operational thought? And if so, did this mean that scientists were operationally evaluating and defining those basic concepts that Stallo, as a misinformed layman, rejected as vague and metaphysical? These are the questions to bear in mind as we examine the intellectual preferences and leanings of American physicists in the years around 1880.

THE RANGE OF

ATOMO-MECHANICAL OUTLOOKS

Harvard's John Trowbridge, late 1870s. (Courtesy of Harvard University Archives)

4

THE ORTHODOXY OF MAYER AND DOLBEAR

A cursory survey of the intellectual leanings of American physicists in the decades bracketing 1880 reveals two basic viewpoints: a large number of physical scientists held mechanical outlooks and spoke exclusively in a language of matter and motion, atoms and ether; but at least a few of them had partially broken from the confines of such mechanical language through qualifications, deviations, or alternatives. A closer inspection shows that even the outlooks of the mechanical scientists were not homogeneous but actually subtly diversified. There were differences on the level of substantive scientific content (actual theories, hypotheses, and models) and the closely related level of scientific ideology (philosophical stance, methodology, and values). For example, one "mechanist" may have coupled a research program based on hard inelastic atoms moving in empty space (the substantive content) to an avowed faith that inductive science is revealing the ultimate truths of nature (the ideology). Another "mechanist" may have coupled a research program based on complex vortex-ring atoms in the ether (the substantive content) to a cautious distrust of the ontological significance of all scientific speculation (the ideology).[1]

The realization that American physical scientists espoused a variety of mechanical as well as potentially nonmechanical outlooks suggests a general response to the questions posed at the end of the last chapter. We can broadly state that Stallo and his critics were focusing on different elements in a complicated matrix of scientific opinion existent among physical scientists. In other words, although atomo-mechanical outlooks dominated, they were not as monolithic and deep-seated as Stallo suggested. But this is a slack and facile response to four originally exact questions. We can gain more detailed answers by examining the actual outlooks regarding content and ideology of representative American physical scientists.

Alfred Mayer and Amos Dolbear, the oldest and most orthodox mechanists to be discussed, shared outlooks that resembled Stallo's stereotype. Born in 1836 and 1837, Mayer and Dolbear were, in a general sense, metaphysically committed to atomo-mechanical physics. Having launched their careers in physics around the decade of the 1860s—a period of unsettled educational and professional patterns in American physics—they picked up perspectives that reflected the prevailing mechanical fashion advocated by European physicists like Thomson and Tait. Mayer and Dolbear's American successors in physics would increasingly have wider and more formal schooling as well as broader research horizons, particularly in the expanding field of electromagnetism. Consequently, we will find these later mechanists to be more diversified in outlook and more difficult to pigeonhole than Mayer and Dolbear.

Alfred Mayer (1836–1897) largely taught himself physics. In fact, even though he was from a wealthy Baltimore family, he received essentially no formal schooling as a youth. The first lecture on physics he ever heard was reportedly one he himself delivered at age twenty-one.[2] Mayer did have, however, an exceptional physics tutor—Joseph Henry, the nation's leading physicist, who also happened to be one of his father's social acquaintances. Soon after Henry's death in 1878, Mayer spoke affectionately about his lifelong mentor: "Professor Henry . . . was the friend of my boyhood; and during 25 years he ever regarded me—as was his wont to say—with 'a paternal interest.' To his disinterested kindness and wise counsels is due much, very much, of whatever usefulness there is in me."[3] With Henry's help, in the late 1850s and early 1860s, Mayer obtained his first college teaching positions. During the Civil War he went abroad for two years, receiving his only formal physics training as a student under Henri Regnault at the University of Paris. Returning to the United States, he held a succession of teaching positions until accepting, in 1871, at age thirty-five, a permanent professorship at the newly founded Stevens Institute of Technology in Hoboken, New Jersey. His first decade of teaching at Stevens Institute was also his most fertile period of scientific research. During the 1870s, he achieved national prominence for his studies on acoustics and, to a slightly lesser degree, for his work on magnetism, electricity, heat, and light. Although weak in mathematics and primarily an experimenter, Mayer did have general theoretical interests. And, like Henry, he eschewed technological applications of his scientific investigations.

[36]

In 1868, three years prior to joining Stevens Institute, Mayer published his *Lecture-Notes on Physics,*[4] a 112-page outline of a comprehensive course developed while teaching at Pennsylvania College, Gettysburg, and at Lehigh University. Mayer revealed in the *Lecture-Notes* a decided atomo-mechanical orientation: he endorsed the premise that all physical phenomena are ultimately reducible to matter in motion and the theory that all matter is ultimately reducible to atoms. He began his *Lecture-Notes,* for example, with a series of "Definitions," the first of which was: "Physical science is the knowledge of the laws of the phenomena of matter." He added that "according to the atomic theory . . . all matter consists of exceedingly minute, absolutely hard and unchangeable *atoms.*" Moreover, "every physical phenomenon is either motion or the result of motion." Examples of the last assertion existed, according to Mayer, "in the phenomena of astronomy, mechanics, acoustics, light, heat, electricity, chemistry, botany, [and] zoology." He did not amplify on the atomo-mechanical aspects of botany and zoology, except to repeat the usual precept of nineteenth-century mechanical reductionism that all bodily senses, including sight, ultimately involved matter in motion. "In the cases of sound, light, and of heat, these effects are propagated through an intervening elastic medium whose particles, vibrating in unison with the particles of the sounding, luminous, or heated body, cause the nerves of the ear, of the eye, or of the skin to pulsate." Here, regarding the transmission of light and heat, Mayer introduced the further mechanical notion of the ether, which he formally defined as a "highly elastic and rare medium" permeating interplanetary as well as interatomic space.

Although the axioms of Mayer's atomic theory did not exactly match the four axioms enumerated and attacked by Stallo, they did come close to at least three—those involving the equality, hardness, and inertness of atoms. As for the fourth axiom, that all energy is kinetic, Mayer deviated from Stallo's preconception by freely employing the concept of potential energy. This discrepancy, however, probably occurred because the purely dynamical physics of P. G. Tait and his British colleagues was not widely disseminated by 1868.

In a later section of the *Lecture-Notes,* Mayer elaborated on his version of atomic theory.[5] Whereas he had earlier described simple

[37]

Alfred Mayer. (Smithsonian Institution)

billiard-ball atoms, he subsequently refined this basic, timeworn concept by adding the eighteenth-century insights of Roger Boscovich. More accurately, he espoused Boscovich's atomic theory as revised and extended by "the researches of Dalton, Joule, Thomson, Faraday, Tyndall, and others." "This theory together with the doctrine of the Conservation of Energy," he insisted, "are the two most important generalizations in Physics." Mayer listed the updated axioms of the Boscovichean theory. Instead of Boscovich's immaterial, pure-force, point "atoms," he retained atoms that were "indefinitely small but finite . . . of extreme hardness . . . indivisible, and unalterable . . . and endued with impenetrability and inertia." To this traditional view he merely tacked on Boscovich's insight, postulating that the billiard-ball atoms were endowed with varying interatomic attractions and repulsions as described by the eighteenth-century natural philosopher.

Although he introduced the concept of elastic ether in his *Lecture-Notes,* Mayer did not dwell on it. In a presentation at Yale four years later, however, he expanded on this and related issues.[6] In this 1872 speech, titled "The Earth a Great Magnet," he demonstrated an electromagnet's ability to induce a current in a coil over long distances. Like Faraday, he rejected the idea that the induced current resulted from magnetic action at a distance, bolstering his opinion with Newton's statement on the "absurdity" of gravitational action at a distance. Whether the all-pervasive ether was acted on by the electromagnet or whether the magnet actually emanated something "immaterial or material," Mayer was not sure. In performing the induction demonstration, nevertheless, he was certain that "the whole space enclosed by the walls of this house contains—is permeated with—something. It goes through your clothes, it penetrates your bodies and saturates your brains. It *must* do so; something must be around us and within us, for surely out of *nothing* I cannot evolve *something*—a current of electricity; and a current of electricity is surely something." This denial of action at a distance but earlier espousal of noncontact, Boscovichean, interatomic forces, would have led Stallo to conclude that Mayer was logically inconsistent in his physics.

Shifting our analysis from the realm of substantive content— matter in motion and atoms and ether—to the realm of ideology, we find that Mayer reflected Joseph Henry's Victorian faith in physical science, natural law, and inductive methodology. Mayer believed in a God-given order in nature and in "constant" and "universal" natural laws.[7] He also believed that science was making real progress in penetrating ultimate truths. Thus, in closing the introductory chapter of his *Lecture-Notes,* he quoted Henry as saying that the study of the physical sciences "habituates the mind to the contemplation and discovery of truth. It unfolds the magnificence, the order, and the beauty of the material universe, and affords striking proofs of the beneficence, the wisdom, and power of the Creator." This deistic leaning persisted in Mayer's thought. A decade later, in 1879, he concluded his popular and pedagogically innovative book, *Light: Simple, Entertaining, and Inexpensive Experiments,* by paraphrasing the 1868 quotation from Henry: "Should you learn nothing else [by making experiments on light], you will see for yourself with what skill, wisdom, and goodness, all these beneficent laws have been arranged. These things come not by chance, or of themselves. They all point to a great and wise Creator, who has given the light a pathway, and filled it with bewildering and perpetual beauty." In like manner, he closed his 1872 speech at Yale by hoping that he had provided his audience with "an insight into those methods by which men of science work out great truths; and *Truth,* is of all value *in itself,* simply because it is truth; irrespective of any practical application it may contain."[8] Such talk of transcendent "truths" and "beneficent laws," when considered alongside Mayer's engrossment in mechanical physics, would probably have led Stallo to judge that Mayer was metaphysically committed to an atomo-mechanical outlook. Indeed, this is precisely what Stallo concluded about the caustic reviewer of his book in *The Critic,* a reviewer who was, most likely, Mayer.

Mayer merged this faith in science and natural law with an enthusiasm for inductive methodology. In the introductory chapter of his *Lecture-Notes,* he quoted lengthy discussions on induction and Baconian philosophy given by leading nineteenth-century thinkers. Specifically, he reprinted statements by Joseph Henry as well as John Stuart Mill. He also quoted from William Thomson and P. G. Tait's famous 1867 text, *Treatise on Natural Philosophy.* In presenting this synopsis of views on the role of "experience"

[40]

and "particular fact" in science, Mayer revealed at least an awareness of the philosophical subtleties of inductive thinking. Not only did he carefully differentiate between proven theories and tentative hypotheses, but he went on to quote Henry's warning that all so-called proven laws and theories are actually "contingent truths." Despite his mention of this caveat, Mayer himself continued to speak of "true and complete" theories, of which "the theory of gravitation and the undulatory theory of light affords us beautiful and instructive examples."[9] In other words, while aware of basic methodological snags and snarls, he pressed onward in his science with optimism and faith that mechanical physics was providing insight into the God-given order of nature.

Mayer exhibited his faith in mechanical physics not only in pedagogical and popular utterances but also in technical scientific papers. That is, his mechanical doctrines were not merely instructional simplifications but actual guideposts for research. During 1871 in the *American Journal of Science and Arts,* for example, he reported "certain experiments of precision" involving induced currents in wire loops to test Auguste de la Rive's mechanical explanation of Faraday's "electro-tonic state." Although the experiments were inconclusive, Mayer judged de la Rive's particulate hypothesis to be inferior to one recently proposed by Yale professor William A. Norton.[10] Around 1865, Norton had developed a "Molecular Physics" involving such concepts as "ethereal pulses" and atoms made from "the condensed universal ether." Characteristically, whereas Mayer favored Norton's "Molecular Physics," Stallo was critical of this same theory, particularly Norton's supposition of a separate electric as well as luminiferous ether.[11]

Atomo-mechanical premises also permeated the best-remembered of Mayer's experimental studies—his 1878 work on the configurations of floating magnets (needles in corks) exposed to an overhead magnet. When, in 1903, during his Silliman Lectures at Yale, J. J. Thomson brought prominence to Mayer's floating magnets by adopting them as a model for electrons in atoms,[12] he actually was resurrecting and updating an old idea. Specifically, Mayer had originally introduced his floating magnets to help explain and illustrate "the action of atomic forces, and the atomic arrangement in molecules." Soon after announcing his findings, he was "much gratified" to learn that William Thomson agreed that the arrange-

Amos Dolbear. (Tufts University Archives)

ments of floating magnets had basic implications for atomic theory. Thomson wrote in *Nature* that Mayer's experiment was "of vital importance in the theory of vortex atoms" and, further, gave "a perfect mechanical illustration . . . of the kinetic equilibrium of groups of columnar vortices revolving in circles round their common center of gravity."[13]

In surveying the molecular phenomena illustrated by various configurations of rings and "nuclei" of magnets, Mayer was careful to stress the hypothetical nature of such models. Nevertheless, he approached this research with his usual optimism and trust in natural law. "These experiments with floating magnets," he wrote in the final paragraph of his main article, "give forcible presentations of the reign of law. It is indeed quite impressive to see order being evolved out of chaos as we hold a magnet over a number of needles, carelessly thrown on water, and witness them . . . entering into the structure of that geometric figure which conforms to the number of magnets composing it."[14] This was the same self-confident scientist who, when confronted by Stallo's 1882 assault on such cornerstones of physics as the undulatory theory of light and the kinetic theory of gases, asked with rhetorical disbelief: "Is it expected that physicists will at once abandon the carefully-framed vessels on which they have sailed to discoveries so triumphant, with not even a raft prepared on which to take their chances?"[15]

One of Mayer's contemporaries who would never abandon the Victorian vessel—even when it began visibly to list toward the end of the century—was Amos Dolbear (1837–1910). In fact, through the 1890s, this physics professor from Tufts College blithely scurried around the vessel, eagerly adding a web of atomo-mechanical riggings to its increasingly top-heavy superstructure.

Dolbear's initiation into physics had been indirect and gradual, following the usual pattern of his American contemporaries.[16] Spending his younger years as a manufacturer's apprentice and a school teacher, he eventually attended Ohio Wesleyan University, graduating in 1866 at the age of twenty-nine. During the next few years, he combined teaching science at various colleges with studying engineering at the University of Michigan. Finally, in 1874 he accepted a permanent position as professor of physics and astronomy at Tufts College in Massachusetts. Unlike Mayer, Dolbear was engrossed in the technical applications of physics; in fact, during the late 1870s and 80s, he frequently engaged in legal battles with companies like Bell and Western Union over patent rights for telephone devices. Dolbear contributed to the basic physics of his day primarily as an educator and popularizer. Perhaps the book that reveals most clearly his perception of contemporary physics was *Matter, Ether, and Motion.*[17] Though published in 1892, more than twenty years after Mayer's *Lecture-Notes,* Dolbear's book betrays that he was of an age with Mayer.

Dolbear's atomo-mechanical preconceptions in *Matter, Ether, and Motion* were so stark and simplistic that Stallo, had he read the book, would probably have deemed it a parody. Dolbear argued that contemporary science was a unity, in contrast to the diversity of early nineteenth-century science with its proliferation of imponderable "fluids" and distinct "forces." All phenomena, physical as well as biological, were "reducible" to—as the book's title implied—matter, ether, and motion. With regard to the categories of matter and motion, Dolbear was adamant that there was no conceptual difference between visible and atomic motions. It was manifest that atomic as well as macroscopic phenomena existed and both were "reducible to the principles of mechanics." If a person doubted this verity and spoke of atomic motions "as if they were mysterious," then it was obvious that this person "cannot have strong mechanical aptitudes, and is not gifted with an adequate scientific imagination." Since Dolbear further believed that "motion was involved in every case where physical energy was involved," he discounted potential energy and concluded that all energy was kinetic. "This view," he added, "is now held by those who have taken the pains to think out the necessary relations that are involved in this subject." As an example of a scientist who had taken the pains, he cited P. G. Tait—the same leading advocate of purely dynamical physics whom Stallo had criticized a decade earlier.[18]

Having explicated two of the terms in the title of his book, matter and motion, Dolbear went on to the third, ether. Dismissing the thought of action at a distance as "not rational," Dolbear insisted on the existence of one multipurpose ether. This ether transmitted light, magnetism, electricity, gravitation, and perhaps even thoughts, as suggested in Stewart and Tait's *Unseen Universe* (another of Stallo's earlier targets). And although Dolbear acknowledged that some physicists regarded the ether merely "as a convenient working hypothesis," he maintained that "this is not the attitude of philosophic minds." Confident of the ether, Dolbear was enthralled by William Thomson's idea of vortex-ring atoms formed out of the pervasive medium. Granting that the vortex theory represented inferential and not demonstrated knowledge, Dolbear nevertheless insisted "it is either that theory or nothing. There is no other one that has any degree of probability at all. If what is presented herewith is not the precise truth concerning a most difficult subject, it may have the merit of helping one to conceive the possibilities there may be of deducing qualities from motion. . . ."[19]

A final phenomenon or "quality" that Dolbear presumed deducible from "motions" was life. Biologists now took for granted, he contended in words recalling those of Du Bois-Reymond, "that vital force as an entity has no existence, and that all physiological phenomena whatever can be accounted for without going beyond the bounds of physical and chemical science." Researchers studying protoplasm had recently shown that "its qualities are the expression of the various movements, chemical and physical, and belong to it simply as a chemical substance." "If such be the case," Dolbear surmised, "it is clear that the solution of every ultimate question in biology is to be found only in physics, for it is the province of physics to discover the antecedents as well as the consequents of all modes of motion."[20] Dolbear, even more than Mayer, represented an extreme in American scientific thought. His firm trust in the mechanical unity of the physical and biological universe fits well Stallo's image of a metaphysical commitment to atomo-mechanical physics.

[45]

5

TROWBRIDGE AND ROWLAND: CAUTIOUS MECHANISTS

Harvard professor John Trowbridge, like Mayer and Dolbear, was deeply immersed in atomo-mechanical physics. So too was Trowbridge's colleague at Johns Hopkins, Henry Rowland. But in contrast to Mayer and Dolbear who were slightly older, Trowbridge and Rowland were more cautious and guarded in their mechanical outlooks. In Trowbridge's case, this caution translated into a skepticism of ultimate answers in mechanical science. For Rowland, the caution meant a wariness of detailed atomic models. Compared with Mayer and Dolbear, Trowbridge and Rowland were also more involved in the latest research on electricity, magnetism, and light; these younger Americans reflected more clearly the shifting focus of international physics toward electromagnetic and optical studies and away from the kinetic theory of gases.

Trowbridge and Rowland revealed their cautious attitudes in a number of writings and speeches, but nowhere more lucidly than in their presentations to the large gathering of physicists in Philadelphia during 1884. In early September of that year, many important American and British physicists converged in Philadelphia for three concurrent and related events: the National Electrical Conference, the International Electrical Exposition, and the annual meeting of the American Association for the Advancement of Science (AAAS). The latter meeting alone drew a total of over 900 American and 300 British scientists. Henry Rowland was president of the Electrical Conference and Sir William Thomson was vice-president; in the AAAS, John Trowbridge was vice-president in charge of Section B–Physics. American participants in the Conference included Simon Newcomb, J. Willard Gibbs, and George Barker, to name only a few. Among the British "Conferees" were James Dewar, Francis Fitzgerald, George Forbes, Oliver Lodge, Arthur Schuster, and Silvanus Thompson. The British scientists, like many of their American colleagues, had come to Philadelphia from the Montreal meeting of the British Association for the Advancement of Science (BAAS), held during late August and chaired by Lord Rayleigh. Moreover, some of these same American and British physicists went on from Philadelphia in early October to William Thomson's famous "Baltimore Lectures"; Lord Rayleigh as well as Rowland and younger Americans like Albert Michelson and Henry Crew were all in his audience. For American physicists, the year 1884 was a high spot in assembly and discourse, not to be matched again until the 1904 St. Louis Congress of Arts and Science.[1]

Trowbridge's contemporaries agreed that he was an exceptionally amiable and kind man. With tongue in cheek, William James once wrote to him: "Though the Lord seems to have withheld from you his crowning grace, of making you a believer in telepathy, he has done the next best thing by you, and created you 'a perfect gentleman.'" Trowbridge (1843–1923), grew up in a wealthy Boston family.[2] In contrast to Mayer and Dolbear, his entrance into physics was smooth and direct, foreshadowing the coming professionalism in American physics. He attended Harvard's Lawrence Scientific School, studying physics under Joseph Lovering and graduating in 1865 at the age of twenty-two. After a few years of teaching mathematics at Harvard and physics at the Massachusetts Institute of Technology, he became in 1870 an assistant professor of physics at Harvard and in 1888 Director of the recently opened Jefferson Physical Laboratory. Throughout his Harvard career, Trowbridge helped break new educational ground: he advocated the laboratory method of teaching physics and encouraged original research by advanced students. In the words of Edwin Hall, one of Trowbridge's younger Harvard colleagues, "the modern spirit entered the Physics Department" when Trowbridge joined the faculty. Trowbridge himself was an avid researcher, particularly in electricity and magnetism. From the 1870s on, he published numerous papers in the *American Journal of Science,* of which he was an associate editor, and the *Proceedings of the American Academy of Arts and Sciences.* His papers were usually reprinted in England in the *Philosophical Magazine.* Perhaps he exerted his greatest influence on American physics through the many graduate students he helped train. Charles St. John, for example, went from Harvard in 1897 to head the Oberlin College physics department; he took with him to Ohio the latest views on theory and experiment as well as "the newer ideals in the teaching of science."[3]

[47]

Trowbridge succinctly expressed his intellectual preferences in the speech he made at the September 1884 meeting of the AAAS— the meeting held in conjunction with the Philadelphia Electrical Exposition and the National Electrical Conference. The speech was appropriately titled, "What is Electricity?" He began by bluntly declaring that he did not know nor would any scientist ever know the true nature of electricity. In posing the question, "What is Electricity?," he was merely encouraging his audience to ask the question "with more humility, and a greater consciousness of ignorance"; such an admission of ignorance signified a "more learned" person. While insisting that nature's ultimate secrets were forever beyond the probings of scientists, Trowbridge nevertheless went on to express optimism and enthusiasm about the future of science. He rhapsodized repeatedly, for instance, about "the great promised land which lies before us." "It is the duty of the idealist to point out the way," he commented, "to greater progress and to greater intellectual grasp." Specifically, such progress was to come by using the principle of conservation of energy to uncover relationships and connections between phenomena. Even though Trowbridge was convinced "that we shall never know what electricity is, any more than we shall know what energy is," he was equally convinced that scientists would be able to discover "the relationship between electricity, magnetism, light, heat, gravitation, and the attracting force which manifests itself in chemical changes." In seeking such connections, however, Trowbridge was not espousing a form of thermodynamic phenomenalism or positivism, because he felt that these energy connections were at root atomo-mechanical.[4]

With one of his typical Biblical allusions, Trowbridge explained that "we have one great guiding principle which, like the pillar of cloud by day, and the pillar of fire by night, will conduct us, as Moses and the Israelites were once conducted, to an eminence from which we can survey the promised scientific future. That principle is the conservation of energy." Unlike their predecessors fifty years earlier, physicists of the 1880s no longer invented "different kinds of forces," electric, chemical, and vital, to explain different types of phenomena; modern physicists, instead, sought to unify their science through energy conservation. In the tradition of Joule, Helmholtz, and their followers, Trowbridge argued that because electrical and magnetic energy were convertible into heat, and heat had its equivalent in mechanical work, then electricity and magnetism were reducible to the purely mechanical properties of underlying atoms. "We look for a treatise on physics," Trowbridge declared, "which shall be entitled 'Mechanical Philosophy,' in which all the phenomena of radiant energy, together with the phenomena of energy, which we entitle electricity and magnetism, shall be discussed from the point of view of mechanics." Although scientists already had authored books on the "mechanical theory" of circumscribed fields such as electricity and magnetism, Trowbridge trusted that "what we are to have in the future is a treatise which will show the mechanical relations of gravitation, of so-called chemical attracting force and electrical attracting force, and the manifestations of what we call radiant energy." Such a synthesis merited the designation, "The New Physics"—the title Trowbridge assigned to his innovative, laboratory-oriented, secondary-school textbook published in 1884, the same year as his AAAS speech.[5]

Narrowing the focus to his designated topic, electricity, Trowbridge advised that new knowledge of electricity and its relationship to magnetism, light, heat, gravity, and chemistry was to come "from the advance in the study of molecular physics." Someday, in fact, physicists might have a simple mechanical theory of electricity based on, as Maxwell likewise had envisioned, only "the laws of vortices as axioms."[6] For the time being, however, Trowbridge set aside such speculations on specific atomic models in favor of the more immediate task of comprehending "the motions of aggregations of atoms." Accordingly, he enunciated a basic working hypothesis: that "a difference of electrical potential" is always associated with either the breakup of "the state of aggregation of particles" or the modification of "the force of attraction between masses or molecules." By designing his electrical experiments with this hypothesis in mind, he hoped to explain phenomena like thermo-electricity as well as the "superficial energy" manifested at the interface of different alloys. He hoped also to resolve certain standing problems in physics—those involving specific heats, for example—the same types of problems on which Stallo had concentrated two years earlier. "Is it not reasonable to suppose," Trowbridge asked, "that certain anomalies which we now find in the determinations of specific heats of complicated aggregation of molecules are due to our failure to estimate the electrical equivalent of the movements and interchanges of the molecules?'' Finally, his working hypothesis regarding a link between electricity and atomic motions or dissociations encouraged him to refine some of Faraday's unsuccessful experiments to detect a relationship between electricity and gravitation.[7]

Trowbridge approached his electrical researches self-effacingly, viewing himself as a Baconian collector of empirical data in a science that grows slowly and inductively. Citing the great number of raw astronomical measurements that preceded Newton's theoretical breakthrough, he asked: ''Is it not fully as important that, in our physical laboratories, we should organize our routine work, and provide some great generalizer, like Sir Isaac Newton, with sufficient data of electro-motive force . . . in order that the relations between this energy and the ultimate motion of the molecular worlds may become better known to us?'' Thus, regarding his own ongoing research on the ''superficial energy'' between alloys, he conceded, ''if the work is not brilliant, I hope that it will result in the accumulation of data for future generations.'' Similarly, regarding his molecular studies on thermo-electricity, he was pleased with even negative experimental outcomes: ''I think there is a great field here—in which a large crop of negative results can be reaped— but these negative results I can not regard entirely as thistles.'' Rather than being useless, such negative results helped physicists inch forward in their science by eliminating certain once-imaginable relationships.[8] To sum up, Trowbridge was a modest mechanist. He was convinced that nature was an atomo-mechanical unity, but was equally certain that physicists would never obtain ultimate answers. And in his own research, he was content merely to contribute to the store of minor empirical findings.

The brilliance of Henry Rowland (1848–1901) was attested to by his close colleague, John Trowbridge. Writing in 1882 from Europe, where he and Rowland were representing the United States in the International Commission of Electricians, Trowbridge raved about the enthusiastic reception of Rowland's work on spectra and diffraction gratings by the French, German, and particularly, British scientists. Trowbridge exclaimed that Rowland, then thirty-four, had ''already made himself immortal.'' This appraisal of

[50]

Henry Rowland. (Smithsonian Institution Photo No. 55802)

Rowland's stature was reinforced two years later by Lord Rayleigh (John W. Strutt) as he traveled through the eastern United States after the Montreal meeting of the BAAS. His 1884 trip agenda with its list of persons whom he visited read like a "Who's Who in American Physical Science": Simon Newcomb, Lewis Rutherfurd, Albert Michelson, George Barker, Charles Pickering, and John Trowbridge, to name a few. From this impressive group, the British scientist singled out one man as being "about the first physicist in America"; this was "Rowland of the Baltimore University." Recollections by one of Rowland's earliest students at Johns Hopkins further reinforced this appraisal of Rowland's prominence. When Robert Millikan mentioned in an obituary of his mentor, Albert Michelson, that Michelson had become, by 1880, the "best known American physicist," Edwin Hall wrote a letter to *Science* upbraiding the younger Millikan, pointing out that the physicist with the highest-ranking achievements was, at that time, his own teacher, Henry Rowland.[9]

What were these achievements? Rowland began his ascent into the leadership of physics soon after graduating in 1870 with a degree in civil engineering from Rensselaer Polytechnic Institute.[10] By 1873, after intense private study of Faraday's works, he published in the *Philosophical Magazine* a paper giving a magnetic analogue of Ohm's law. The paper appeared in this prestigious British journal at the instigation of no less a figure than James Clerk Maxwell, to whom the unknown Rowland had sent a copy; ironically, the inexpert editors of the *American Journal of Science* earlier had rejected the same paper. Daniel C. Gilman, president of the newly founded Johns Hopkins University, learned of the prodigious young American and hired him in 1875 to fill the physics position in the faculty he was assembling. Gilman immediately sent Rowland to Europe to find out more about organizing a university laboratory. While there, Rowland visited Maxwell in Scotland and Hermann von Helmholtz in Germany. Using Helmholtz's laboratory facilities, he completed his next major research project, a demonstration of the magnetic effect of moving electrical charges. (Three years later, the young American wrote an indignant letter to the *Philosophical Magazine* complaining that this experiment, entirely his own, was being "constantly referred to as Helmholtz's experiment."[11]) Rowland soon established a well-equipped laboratory at Johns Hopkins and began to train the first of many promising graduate students such as Edwin Hall. He also began his measurements of the mechanical equivalent of heat, which, when published in 1880, brought him acclaim. His successful manufacture of diffraction gratings that had unparalleled dispersive power, along with his associated spectral studies, assured his scientific reputation. It was not all hyperbole when Trowbridge characterized Rowland's 1882 visits to France and England as "Rowland's great triumphant march."[12]

At the apex of Rowland's career, his outlook on physics was a mixture of resignation and hope, doubt and belief. On the one hand, he insisted that most of the basic questions in physics were both unanswered and presently unanswerable. On the other, he trusted that these deeper questions would ultimately be resolved. And whereas he cautiously avoided mention of detailed atomic mechanisms, he structured his thoughts around a broad atomo-mechanical framework. These different strains of Rowland's outlook appear on the most general level in his famous speech, "A Plea for Pure Science," delivered to the AAAS in 1883:

> there will be those in the future who will study nature from pure love, and for them higher prizes than any yet obtained are waiting. We have but yet commenced our pursuit of science, and stand upon the threshold wondering what there is within. We explain the motion of the planets by the law of gravitation; but who will explain how two bodies, millions of miles apart, tend to go toward each other with a certain force? We now weigh and measure electricity and electric current with as much ease as ordinary matter, yet have we made any approach to an explanation of the phenomenon of electricity? Light is an undulatory motion, and yet do we know what it is that undulates? Heat is motion, yet do we know what it is that moves? Ordinary matter is a common substance, and yet who shall fathom the mystery of its internal constitution?
>
> There is room for all in the work, and the race has but commenced.[13]

This litany of unanswered but answerable questions became a common refrain in Rowland's writings and speeches.[14] That he generally anticipated answers to these individual questions in terms of atoms and ether is more evident in one of his later speeches. In 1889, he stated that light was an undulation of the assuredly exis-

[53]

tent ether; he was uncertain only of the "constitution" of this medium. And regarding the internal mysteries of matter, he was confident that "the atoms of matter can vibrate with purer tones than the most perfect piano"; the modern scientist merely could not "conceive of their constitution."[15]

This later speech also helps illustrate his twofold view of scientific knowledge: the admission of momentary ignorance but confidence in the future. After recounting the unanswered questions in physics, he further wrote in 1889: "The proper attitude of the mind is one of resignation toward that which it is impossible for us to know at present and of earnest striving to help in the advance of our science, which shall finally allow us to answer all these questions."[16] Adhering to his principle of resignation, Rowland avoided detailed atomic speculations in his research papers on both spectra and the mechanical theory of heat. "Every equation of thermodynamics proper is an equation between mechanical energy and heat," he wrote in his lengthy report on the redetermination of the mechanical equivalent of heat; the kinetic theory of atomic motions simply did not figure in his discussions of heat and energy.[17] Only in his electrical researches—as we shall see—did he venture an occasional, guarded comment on atomic processes.

But resignation was only half of Rowland's twofold view of scientific knowledge; he also had hope for the future. He coupled his humility before nature with a Victorian pride in the ability of scientists to grasp the laws of nature. Through the 1880s, he extolled the "true and overwhelming progress of science which marches forward to the understanding of the universe," which grows steadily toward "a perfect whole," and which will "finally arrive at a solution" of fundamental problems. There were "exact laws of nature" that scientists could discover while engaged in their pursuit of "truth," and there was a "standard of absolute truth" that existed in the empirical realm of experiment and observation.[18] In his "Plea for Pure Science," he used the metaphor of the frontier to communicate to his American audience his optimistic image of advancing science: "Pure science is the pioneer who must not hover about cities and civilized countries, but must strike into unknown forests, and climb the hitherto inaccessible mountains which lead to and command a view of the promised land—the land which science promises us in the future; which shall not only flow with milk and honey, but shall give us a better and more glorious idea of this wonderful universe."[19]

Only in his writings on electricity and magnetism did Rowland venture to the edge of the frontier and provide glimpses of an atomo-mechanical promised land. He afforded glimpses, for example, in his 1884 presidential address before the Electrical Conference in Philadelphia. Although the Conference was convened in part to establish electrical standards, and although Rowland had carried out important measurements in this industrially critical field,[20] he made it clear in his presidential address that he envisioned a grander role for electrical science. "Shall we be contented with a simple measurement of that of which we know nothing? I think nobody would care to stop at this point, although he might be forced to do so. The mind of man is of a nobler cast, and seeks knowledge for itself alone." The groundwork for this nobler search, Rowland observed, had already been laid by researchers like Poisson, Green, Helmholtz, Thomson, and Maxwell. Rowland himself had purchased a copy of Maxwell's *Treatise on Electricity and Magnetism* when it first appeared in 1873, and he began using Maxwellian ideas within the next few years. By the time of his presidential address, he had advanced well beyond the simple, non-mathematical analogies and models that undergirded his initial studies of electricity and magnetism. He had become a supporter of Maxwell's dynamical theory as presented in the *Treatise.* In this theory, Maxwell deemphasized his earlier, detailed mechanical models in favor of generalized mechanical principles—principles that he had derived from Thomson and Tait's formulation of the analytic dynamics of Joseph Lagrange and others.[21] Rowland stressed that Maxwell and his colleagues had devised a "mathematical" theory of electricity and magnetism. In Rowland's judgment, the theory was devoid of any "hypothesis as to the nature of electricity" and dependent on only "the most simple laws of electricity and magnetism." Built on basic empirical facts, the theory could "never be overturned, whatever the fate of the so-called electric fluid or the ultimate theory of magnetism." In endorsing this "mathematical" theory, Rowland perceived himself to be aloof from all speculation, but he, like Maxwell, harbored

[55]

atomo-mechanical predilections. After all, even the analytic, mathematical theory of electromagnetism was still a dynamical theory, one anchored to classical laws of motion.

There were more explicit hints of Rowland's atomo-mechanical predilections in the remainder of his presidential address. In particular, he pointed out that the electromagnetic theory had altered scientists' ideas about the "mechanism" of electric and magnetic attraction. As a result of Faraday's and Maxwell's development of the notion of lines of force (elastic strains in the ether associated with electromagnetic forces), scientists were now "at liberty to deny the existence of all action at a distance, and attribute it to the intervening medium. . . ."[22] Rowland expanded on this perspective later in the 1880s. In 1889, he contended that the ability of a current in a wire to affect a distant magnetic needle was "evidence that the properties of the space around the wire have been altered rather than that the wire acts on the magnet from a distance." To more fully picture "the property which the presence of the electric current has conferred on the luminiferous ether," he suggested, like Faraday, that lines of force be thought of as "rubber bands, which tend to shorten in the direction of their length and repel each other sideways."[23]

During 1888, Rowland went beyond this illustrative rubber-band model and mentioned a more complex vortex model for electromagnetism: "Thomson and Maxwell have shown that the medium around a wire carrying an electric current is in motion, and that the vortex filaments form Faraday's lines of magnetic force; for Faraday's discovery of the magnetic rotation of the plane of polarization of light can be explained in no other way."[24] While addressing the Electrical Conference in 1884, Rowland had spoken more cautiously about this "vortex motion in a fluid" and the "mechanical model of such a medium." When scientists broach these topics, he warned: "We are face to face with the great problem of nature, and the questions, What is matter? What is electricity? evoke no answer from the wisest among us"; nevertheless, as in his other writings, he trusted that "science will advance and we shall know more."[25]

Rowland's implicit mechanical leanings also surfaced in his 1884 defense of Maxwell's electromagnetic theory as it applied to light. "So perfectly does this theory agree with experiment," he wrote, "that we can almost regard it as a certainty."[26] Consequently, he was disturbed by criticisms of the theory, including one (most likely from William Thomson) which claimed that "it is not a true theory, because it is not mechanical," not reducible to "matter and motion." While professing to be above such speculative and philosophical issues, he carefully explained how the traditional elastic or undulatory theory of light was reconcilable with the more recent electromagnetic viewpoint: "it is to be noted that the old mechanical theory that light is a vibration in a medium having the properties of an elastic solid is not entirely at variance with the new theory. . . . The electro-magnetic theory says that the waves of light are waves of electric displacement, while the old theory says they are waves of ether. Make electricity and the ether equal to each other and the two theories become one." That Rowland himself entertained the possibility of a more general mechanical reconciliation of this type became evident with his frequent declarations of confidence in the existence of the ether—not the conventional elastic-solid ether, of course, but a Maxwellian dynamical ether. He went on in his 1884 address to exclaim, "Truly the idea of a medium is to-day the keystone of electrical theory."[27]

Although fully committed to the central role of the ether in electromagnetic phenomena, Rowland was aware of basic problems with this elusive medium. Perhaps with Michelson's 1881 interferometer experiment in mind, he commented that "it is a most wonderful fact that we have never been able to discover anything on the earth by which our motion through a medium can be directly proved."[28] Such strange results, however, did not shake his commitment. In fact, after Heinrich Hertz in 1887 experimentally confirmed Maxwell's prediction of electromagnetic or "ethereal" waves, Rowland was even more enthusiastic about the ether. In 1889, he stated:

> The luminiferous ether is, to-day, a much more important factor in science than the air we breathe. We are constantly surrounded by the two, and the presence of the air is manifest to us all. . . . The luminiferous ether, on the other hand, eludes all our senses and it is only with the imagination, the eye of the mind, that its presence can be perceived. By its aid in conveying the vibrations we call light, we are enabled to see the world around us, and by its other motions which cause magnetism,

the mariner steers his ship. . . . When we speak in a telephone, the vibrations of the voice are carried forward to the distant point by waves in the luminiferous ether. . . . No longer a feeble, uncertain sort of medium, but a mighty power, extending throughout all space and binding the whole universe together, so that it becomes a living unit in which no one portion can be changed without ultimately involving every other portion.[29]

In addressing the Electrical Conference in 1884, prior to Hertz's detection of ethereal waves, Rowland emphasized another basic member of the mechanical triumvirate of matter, ether, and motion: he assigned a central role to matter as well as to ether. He argued that electricity was neither a fluid nor energy but "a property of matter." Since it lacked inertia and weight, electricity was not matter itself, at least not known forms of matter. It was, nonetheless, always associated with matter; lines of force, for example, always began and ended in matter. Rowland, accordingly, concluded his 1884 discussion of electricity by saying: "The theory of matter . . . includes electricity and magnetism, and hence light; it includes gravitation, heat, and chemical action; it forms the great problem of the universe. When we know what matter is, then the theories of light and heat will also be perfect; then and only then, shall we know what is electricity and what is magnetism."[30]

In summary, despite occasional mention of Maxwellian mechanical models, Rowland consciously shied away from detailed atomo-mechanical explanations, insisting that they were usually beyond the scientists' ken. On the other hand, he implicitly endorsed the broader precepts of mechanical physics, trusting that scientists someday would find ultimate explanations in terms of matter, ether, and motion. If John Trowbridge, with his limited mechanical goals, was a modest mechanist, then Henry Rowland, with his momentary reservations but implicit mechanical hopes, was a muted mechanist.

6

MICHELSON AND HALL:
EXPERIMENTATION AND EDUCATION

Alfred Mayer, Amos Dolbear, John Trowbridge, and Henry Rowland are only representative of the many American physicists who embraced some form of mechanical viewpoint around 1880. Two additional representatives deserve our scrutiny: Albert Michelson and Edwin Hall. Michelson—probably the best remembered American physicist from that period—is worth considering because he is cited so often today for his mechanically inspired experimental program. Edwin Hall, an experimentalist as well as an active educator, is worth examining because his pedagogic writings further illustrate the infusion of mechanical thought into educational doctrines. Michelson and Hall, whose birth dates were 1852 and 1855, are the youngest of the physicists so far surveyed.

Born in Poland but raised in Nevada, Michelson (1852–1931) began formal physics training in 1869 at the United States Naval Academy.[1] After graduating, he returned to Annapolis to teach physical science. With counsel from Simon Newcomb, he there determined the speed of light with unprecedented precision. From 1880 to 1882 he traveled in France and Germany, continuing his studies of light and optics. At the University of Berlin, in Hermann von Helmholtz's laboratory, he carried out his first ether-drift experiment with his newly devised interferometer—an instrument used for comparing the speeds of light beams traveling in different directions. Returning to the United States, he became professor of physics at the newly opened Case School of Applied Science in Cleveland; here, during the mid-1880s, he collaborated with Edward Morley on their more famous ether experiments. Michelson then spent a few years at Clark University, where he continued work that led to a new light-wave method of measuring the standard meter. Finally, in 1892, he became head of the physics department at the recently founded University of Chicago. He organized a superb faculty. Members included George Ellery Hale as well as Robert Millikan, one of Michelson's former graduate students. He

Albert Michelson, ca. 1930s. (AIP Niels Bohr Library, Michelson Collection)

continued to do important research using the spectrometer and interferometer, and in 1899 he helped found the American Physical Society. Lauded for his achievements, he became in 1907 the first American to win a Nobel Prize in science.

Michelson was foremost an experimenter. When occasionally he did make an excursion into the realm of theory, it was always brief and functional, undertaken only to provide support for experiment. Drawing on established fashions in physics, he consequently clothed his experiments in nineteenth-century theoretical garb. This was particularly true of his various ether experiments done in the 1880s. In the first sentence of his first published paper on the ether, he revealed his mechanical predisposition: "The undulatory theory of light assumes the existence of a medium called the ether, whose vibrations produce the phenomena of heat and light, and which is supposed to fill all space." With this general assumption, Michelson agreed. In the next few sentences of this first paper, however, he isolated a particular, still-unproven hypothesis that he sought to test. Early in the nineteenth century, Augustin Fresnel had argued that the ether was essentially at rest relative to the moving earth. If the earth were gliding through this motionless medium, then, as Michelson pointed out, "the time for light to pass from one point to another on the earth's surface, would depend on the direction in which it travels." Desiring to test this proposition, and responding to Maxwell's challenge that the relevant time intervals were too small for measurements, Michelson designed and carried out his 1881 interferometer experiment.[2]

As is well known today in our post-Einsteinian era, Michelson's experimental results contradicted Fresnel's widely held hypothesis that the ether was at rest relative to the moving earth. "The result of the hypothesis of a stationary ether is thus shown to be incorrect," Michelson announced, "and the necessary conclusion follows that the hypothesis is erroneous." This did not mean that the experiment had shaken Michelson's belief in the existence of the ether. In his final paragraphs, he made an oblique reference to George Stokes's 1846 view that the earth might drag along a shroud of locally stationary ether; Michelson was hinting at an alternative hypothesis that might explain his failure to detect an ether wind. In a letter to his financial backer, Alexander Graham Bell, Michelson was less guarded regarding this alternative. After outlining the

[61]

overall theoretical problem and his experimental results, he told Bell: "Thus the question is solved in the negative, showing that the ether in the vicinity of the earth is moving with the earth."[3] Fresnel's theory of the stationary ether was in question, not the ether itself.

In 1884, either at the Montreal BAAS meeting or at Thomson's "Baltimore Lectures," Michelson discussed his ether experiment with Lord Rayleigh and Sir William Thomson. With one of the implications of Fresnel's hypothesis now in doubt, Rayleigh and Thomson encouraged Michelson to repeat Armand Fizeau's earlier confirmation of the other major implication—that the ether enclosed in a moving, optically transparent medium would have a slight speed, thus altering the speed of light in the moving medium. Accordingly, Michelson and his co-worker, Edward Morley, refined Fizeau's experiment on the speed of light in moving water; they reported in 1886 that they had confirmed both Fizeau's prior verification and Fresnel's prediction. Michelson already was looking ahead as early as 1884—when he first considered the Fizeau experiment—to a repetition of his own interferometer experiment.[4] Now, with Fresnel's hypothesis partially vindicated, the need for a repetition was more pressing.

Again, it was Rayleigh who helped instigate a second measurement of the relative earth-ether motion. Earlier distressed by the disinterest of his "scientific friends" and the "slight attention the work received," Michelson confided to Rayleigh in 1887 that, "Your letter has . . . once more fired my enthusiasm and it has decided me to begin the work at once." During the previous year, Hendrik A. Lorentz had published a critique of the first measurement that further motivated Michelson to repeat the experiment. Michelson and Morley in their subsequent and now famous paper, "On the Relative Motion of the Earth and the Luminiferous Ether," acknowledged that they had recently confirmed Fizeau's test of Fresnel's hypothesis regarding the motion of ether in flowing water; they were now retesting the second implication of Fresnel's hypothesis, that the ether was at rest when not in a moving, transparent medium. Their eventual conclusion reaffirmed Michelson's 1881 finding: "It appears, from all that precedes, reasonably certain that if there be any relative motion between the earth and the luminiferous ether, it must be small; quite small enough entirely to refute Fresnel's explanation. . . ."[5]

This new result once again threatened Fresnel's overall theory of the ether but not the ether itself. In fact, as Michelson had done in 1881, he and Morley endeavored to account for their perverse result with alternative explanations. First they reviewed the ether theories of Stokes and Lorentz, pointing out, however, that both of these options had their own weaknesses. Then in a "Supplement" attached to the paper, they rationalized that the earth's irregular surface perhaps had trapped portions of ether thus obscuring the ether wind they aimed to detect: "It is obvious from what has gone before that it would be hopeless to attempt to solve the question of the motion of the solar system by observations of optical phenomena *at the surface of the earth*. But it is not impossible that at even moderate distances above the level of the sea, at the top of an isolated mountain peak, for instance, the relative motion might be perceptible in an apparatus like that used in these experiments."[6]

As he had in 1881, Michelson continued to couple his faith in the luminiferous ether to an orthodox mechanical view of the atomic processes that caused light. Throughout an 1888 address on light experiments, for example, he superficially but confidently spoke of "the forces and motions of vibrating atoms and of the ether which transmits these vibrations in the form of light." In general, the experimental questions Michelson asked of nature during the 1880s were mechanical questions. And the answers he obtained were often unexpected. As one of his American colleagues said in 1889 on the occasion of his receiving the Rumford Medal of the American Academy of Arts and Sciences, "We may thank Professor Michelson not only for what he has established, but also for what he has unsettled."[7]

During the late 1870s, while Michelson was first achieving distinction with his light-velocity experiments, an even younger American, Edwin Hall (1855–1938), was gaining a reputation through his electrical experiments.[8] After graduating from Bowdoin College in 1875, and then teaching for two years in secondary schools, Hall turned to physics. John Trowbridge advised him to avoid the

Edwin Hall, late 1880s. (Courtesy of Harvard University Archives)

Harvard graduate program in favor of Henry Rowland's newly equipped laboratory at Johns Hopkins. While completing his doctoral studies at Johns Hopkins from 1877 to 1879, he discovered, with Rowland's help, the electrical property of conductors soon to become known as the "Hall effect." He then spent a summer in Europe, visiting, among other places, Helmholtz's laboratory in Berlin. In 1881 he took up his lifelong teaching post at Harvard, where he continued to distinguish himself both as a researcher and as an educator with a special interest in secondary schools.

Hall fashioned his experiments within an atomo-mechanical framework that had only a modicum of theoretical subtlety and mathematical refinement. The framework was, however, admittedly provisional and conditional. In later years, Hall's student and subsequent colleague at Harvard, Percy Bridgman, described Hall's simple scientific style: "He was not a mathematician, but he had a very strong physical sense, which was most at home in a mechanistic medium very similar to that of the great English physicists. For him a theory consisted in a painstaking working out by native wits of all the consequences which he could see were inherent in the fundamental physical picture." This scientific style, at least as initially manifested in Hall's doctoral researches on the "Hall effect," also reflected the influence of his mentor, Henry Rowland. Rowland, in his early studies of electricity and magnetism, had depended on concrete models and analogs; later, he adopted Maxwell's dynamical perspective. Hall conducted his doctoral researches within the general context of the Maxwellian perspective while, at the same time, incorporating specific analogs tied to simple mechanisms.[9]

The "Hall effect" refers to a subtle electrical attribute of metals. In more technical language, it refers to the transverse potential difference which is set up across a current-carrying conductor due to a magnetic field acting perpendicular to the current. On a general level, Hall's experimental search for the effect was motivated by Maxwell's earlier theoretical denial of the effect; on a specific level, his search was encouraged by Rowland.[10] Initially, in trying to understand the behavior of an electric current when subjected to a magnetic field, Hall had proposed an analogy involving a stream of water. By 1880, after conclusively detecting the elusive effect, he had devised an alternative mechanical analogy. Before

[65]

introducing this new analogy, he emphasized its tentativeness. "It is perhaps idle to speculate as to the exact manner in which the action between the magnet and the current takes place in any of the preceding experiments," Hall began, "but it may be worth while to remark a seeming analogy, somewhat strained perhaps, between this action and a familiar mechanical phenomenon, the theory of which has of late attracted considerable attention." The familiar mechanical phenomenon was the curved trajectory of a swiftly projected, spinning baseball—a "curve ball" in the vernacular of the rapidly rising, great American pastime. "Imagine now," Hall continued, "an electrical current to consist of particles analogous to the baseball, moving through a metallic conductor . . . . Suppose, further, the particles of electricity, on coming within the influence of the magnet, to acquire a motion of rotation. . . . Under all these supposed conditions we might perhaps expect to find the action which is actually detected." To add credence to his atomo-mechanical supposition that the magnet would cause the electrical particles to rotate, Hall cited a similar supposition in Maxwell's *Treatise on Electricity and Magnetism.*[11]

The Hall effect became a popular topic among physicists during the early 1880s, resulting in reports which were published in leading American and British journals. After hearing Hall describe the phenomenon at the 1881 meeting of the BAAS, Sir William Thomson characterized it as "by far the greatest discovery that has been made in respect to the electrical properties of metals since the times of Faraday. . . ." Hall again received public notice when Thomson presented a paper on the effect to the 1884 meeting of the AAAS.[12] While Hall himself continued to pay only minimal attention to the theoretical intricacies of the effect, he nevertheless suggested through the 1880s that a "molecular" explanation involving "some internal interaction" within metallic conductors should "be regarded as established."[13]

Although Hall was best known among research physicists for this electrical effect, he was most highly recognized in educational circles for introducing into American secondary schools the laboratory method of teaching physics.[14] Joining a movement begun around 1880 by progressive physics educators, among whom were Alfred Mayer and John Trowbridge, and prodded into action by Harvard president Charles Eliot, Hall designed and popularized during the late 1880s a course of experiments for use in secondary schools. The course, structured around the "Harvard Descriptive List of Elementary Physical Experiments," came to dominate secondary-school physics by about 1900. While Hall's primary intention had been to introduce the laboratory method, an additional effect of this endeavor was to reinforce a mechanical perspective on the physical universe.

Hall made clear his mechanical leanings in an article published in 1893 in the *Educational Review* on "Teaching Elementary Physics." In recommending his Harvard course of experiments, he stated:

> The course at present contains forty-six exercises, twenty-one of which are in mechanics, including hydrostatics. This may appear to some a number relatively too large for mechanics, but when we reflect upon the extent to which we are visibly dependent upon mechanical laws, and upon the fact that physicists consider no phenomenon of their science satisfactorily accounted for until it can be explained as a mechanical effect, it will appear that the amount of time devoted to matters of such importance is well spent.

To make this point stronger, he further remarked that even electrical and magnetic phenomena were ultimately atomo-mechanical. Thus, Hall felt that a student who desired to concentrate exclusively on electricity and its practical application should be made aware that "if he ever reaches such an intellectual height as to catch a glimpse of the pioneer thinkers who are working out the *science* of electricity, he will find them engaged with such homely matters as elasticity, friction, and inertia."[15] Perhaps Hall had in mind his own pioneering study of the Hall effect and his homely mechanical analogy between electrical particles and baseballs. In the classroom as well as in the laboratory, Hall worked within a simple mechanical framework. Like Michelson, he was practical and down-to-earth in his commitment to atomo-mechanical precepts.

[67]

7

MESSAGES FROM EUROPE

The careers of Mayer, Dolbear, Trowbridge, Rowland, Michelson, and Hall have shown that American physicists derived many of their mechanical outlooks from their European counterparts; especially influential were Faraday, Thomson, Tait, Maxwell, and Helmholtz. In general, Americans assimilated European mechanical notions by studying overseas, by reading foreign journals and texts, and by interacting with European scientists visiting the United States. To be sure, not all the outlooks transmitted from Europe to America during the 1870s and 1880s were mechanical. Some Continental physicists, Gustav Kirchoff for one, had eschewed the atomo-mechanical view in favor of a more phenomenological approach. At this point, however, our concern is with mechanical thinkers.

In the decades bracketing 1880, American physicists often undertook either graduate or postgraduate study in Europe. In a statistical survey of the American physics community, Daniel Kevles found that, of approximately fifty physicists identifiable as the most productive researchers in the years from 1870 to 1893, about one-third had studied abroad. Kevles concluded that students were induced to study in Europe by the shortcomings of American graduate schools; the doctoral programs at Johns Hopkins, Harvard, and Yale, for example, were still fledgling.[1] American physicists going overseas most often chose to study in Germany, and especially at the prestigious universities of Berlin, Göttingen, Heidelberg, and Leipzig. Even an eminent British physicist encouraged this Germanic pattern of study: John Tyndall, on completing his 1873 lecture tour of the United States, used his tour profits to establish an American scholarship for "the education of young philosophers in Germany."[2]

Americans were particularly attracted to the University of Berlin, the site of Helmholtz's famous laboratory during the 1870s and 1880s; the careers of Rowland, Michelson, and Hall showed this clearly. Americans respected Hermann von Helmholtz (1821–1894)

[68]

not only as one of the discoverers of conservation of energy but also as an innovator in nearly all branches of physics, including biophysics. When, in 1893, the aged professor came to Chicago as honorary president of the International Electric Conference held in conjunction with the World's Fair, he found himself in the appreciative company of many of his ex-students. One observer of the Chicago trip—Helmholtz's final important public appearance—recorded that Helmholtz "aroused much enthusiasm among the younger physicists of the country, many of whom have been his pupils." "While in this country every honor was shown him," another observer recalled. "Here he found many of the hundreds and thousands of his pupils who everywhere in the world are adding luster to his name by perpetuating his spirit and methods. . . ."[3] Generally speaking, the spirit and methods of this revered teacher were mechanistic. He particularly favored those physical analogies and atomic models that had firm mathematical foundations. He viewed these mechanisms, however, as being merely useful conceptual constructions that lacked metaphysical import. According to a recent biographer, Helmholtz displayed "a profound distaste for metaphysics and an undeviating reliance on mathematics and mechanism." For example—as Rowland and Hall surely realized during their stays in Berlin—Helmholtz was more comfortable with the Maxwellian view of electromagnetic action through a medium than with the prevailing Continental electrodynamics as advocated by Wilhelm Weber.[4]

Whereas American physicists studying overseas usually attended a German university, American physicists at home normally read British periodicals and texts. In addition, Kevles has found that the fifty or so most productive American physicists during the years between 1870 to 1893 published over twice as many of their own articles in British rather than in German journals, their favorite foreign outlets being the *Philosophical Magazine* and *Nature*.[5] In other words, even though Americans appreciated the excellence of German science, they usually found it easier in their everyday professional lives to follow and participate in the scientific literature of their native tongue. Germany dominated in the sphere of overseas education, but Britain ruled in the realm of foreign scientific writing and journalism. American physicists reflected their engrossment in British literature not only in their propensity to publish in British journals but also, more generally, in their emulation of past British luminaries and deference to contemporary British authorities. Usually, these role models and authorities had strong mechanical leanings.

If one were to choose the European physicist from times past whom Americans of the 1870s and 1880s most venerated, it would be Englishman Michael Faraday (1791–1867). Alfred Mayer characterized Faraday as "that wonder of experimental fertility" and "our great master of experiment." When referring to Faraday's discovery of electromagnetic induction and explanation of the discovery in terms of lines of force, Mayer added: "With a master's hand he traced the strong broad outlines of the truth, and for forty years philosophers have pondered on these facts and have made thousands of experiments, yet have failed to fathom their full significance." One of these pondering philosophers was John Trowbridge. "I have often reflected upon these experiments of Faraday," Trowbridge wrote while refining Faraday's methods of detecting a connection between electricity and gravitation. The American most directly inspired by Faraday, however, was Henry Rowland. While in college he began an intensive study of Faraday's writings, and had come, by the late 1880s, to "speak his name with reverence." Like many of his American colleagues, Rowland included Faraday's ideas on lines of force "among the foundation stones of the edifice of our science."[6] In emulating Faraday, physicists like Mayer, Trowbridge, and Rowland revealed much about their own scientific outlooks. The Americans were conceptually attuned to the writings of this Englishman—a nonmathematical experimenter, an independent thinker of strong physical intuition, a staunch critic of action at a distance, an advocate of physically real lines of force, and a tireless searcher for interconnections among phenomena.[7]

American deference to contemporary British rather than to German or French scientific authorities was also the norm during the 1870s and 1880s. Although Americans were mindful of Helmholtz's ongoing publications, they were most sensitive and responsive to the recent writings of William Thomson, P. G. Tait, and James Clerk Maxwell. The extent to which the writings of these three individuals served as American gospels of sanctioned scientific thought was illustrated in an 1884 squabble involving Edwin

Hall. Hall wrote a lengthy letter to *Science* aiming to provide physics teachers with a "proper definition" of the hitherto ambiguous term "inertia," precise terminology being an important international issue during this period. To add weight to his interpretation of the term, Hall cited the authoritative views of Maxwell in his *Theory of Heat* (1870) and Thomson and Tait in their widely respected *Treatise on Natural Philosophy* (1867). Within a few days, the editor of *Science* had received five letters critical of Hall; the writers objected not so much to Hall's personal interpretation of "inertia" but to his misinterpretation of Maxwell, Thomson, and Tait. The letters and Hall's rebuttal can be likened to a quarrel among fundamentalist clergymen—rich in specific scriptural allusion, with chapter and verse glibly quoted, but lacking in fresh theological import. For example, to redress the "injustice" of Hall's partial and self-serving quotation from Maxwell's *Theory of Heat,* Thomas Mendenhall cited Maxwell's entire passage regarding inertia. Moreover, Mendenhall facilely traced Maxwell's developing thought on inertia throughout his career beginning with his "On the Properties of Matter" written at seventeen, passing on to his *Matter and Motion* (1877), and even including his review of Thomson and Tait's influential text, *Natural Philosophy.* Similarly, Mendenhall had at his fingertips appropriate passages from Thomson and Tait, "the restorers of Newton," to contrast with Hall's original contentions. Responding to Mendenhall's recitation of chapter and verse—and to similar detailed recitations by physicists Charles Hastings and Alfred Gage—Hall undertook his own close textual analysis of the British scientific gospels. His rebuttal is laced with hair-splitting rejoinders such as: "It was my belief, however, and it still is, that Maxwell, in that famous chapter, used 'mass' in two senses. . . ."[8] In turning mainly to the writings of Maxwell, Thomson, and Tait to arbitrate their dispute, Hall and his adversaries illustrated the Americans' deep immersion in, and dependence on, British publications—in this case, publications notable for their pervasive mechanical messages.

Beyond study abroad and foreign publications, American physicists could also hone their mechanical wits through interactions with European visitors to the United States. Once again, English-speaking physicists hailing from Great Britain had the largest impact during the decades around 1880. The 1893 visit of the aged, ailing, and German-speaking Helmholtz just was not as consequential as the earlier visits of Britishers John Tyndall and William Thomson, who were in their prime. John Tyndall (1820–1893) undertook his lecture tour of major cities in eastern United States during the fall and winter of 1872–73. His American sponsors hoped that he could promote the cause of science among the educated public, since his books, especially *Heat as a Mode of Motion,* were widely admired in the United States. His lectures, all on the topic of light, were geared to a popular audience. Nevertheless, practicing physicists became actively involved in Tyndall's endeavor. Joseph Henry helped schedule the tour, and Alfred Mayer helped plan Tyndall's farewell banquet. This send-off was a lavish gathering at Delmonico's in New York; 200 prominent American educators, publishers, professional men, and scientists attended.[9]

Tyndall increased his American outreach and impact by distributing during his tour a lecture summary that "sold by the hundred thousand," and then later enlarging this summary into a book that went quickly through a number of American editions. Regarding the texts of the lectures, Tyndall stated that except for "a fragment or two, not a line of them was written" before arriving in the United States; moreover, for the sake of persons who missed his programs, he saw to it that the lectures were published swiftly and in "an authentic form."[10] Consequently, Tyndall's book, *Lectures on Light* (1873), was an accurate rendering of the European message American physicists were not only reading but also hearing in the early 1870s. What was this message?

In his lectures, Tyndall provided some justification for Stallo's later assertion that the British scientist and popularizer was "one of the most strenuous advocates of the atomo-mechanical theory." Actually, what he did was join an atomo-mechanical predisposition to a traditional hypothetico-deductive methodology. That is, in conventional nineteenth-century style, atomo-mechanical hypotheses were to be succeeded by logical deductions that in turn were to be empirically tested. A physicist, for example,

> cannot consider, much less answer, the question, "What is light?" without transporting himself to a world which underlies the sensible one, and out of which, in accordance with rigid

law, all optical phenomena spring. To realize this subsensible world, if I may use the term, the mind must possess a certain pictorial power. It has to visualize the invisible. It must be able to form definite images of the things which that subsensible world contains; and to say that, if such or such a state of things exist in that world, then the phenomena which appear in ours must, of necessity, grow out of this state of things. If the picture be correct, the phenomena are accounted for; a physical theory has been enunciated which unites and explains them all.[11]

Tyndall's primary example of such subsensible pictures involved light, which was to be considered as a wave phenomenon analogous to sound. "In the case of light," he explained, "we have in vibrating atoms of the luminous body the originators of the wave-motion, we have in the ether its vehicle, while the optic nerve receives the impression of the luminiferous wave." Elaborating on the ether, he added: "To this interstellar and interatomic medium definite mechanical properties are ascribed, and we deal with it as a body possessed of these properties. . . . We treat the luminiferous ether on mechanical principles." With confidence and even aplomb, Tyndall detailed in his lectures this atomo-mechanical, hypothetico-deductive view of light. He was insistent, as stated in his concluding lecture, that "the roots of phenomena are embedded in a region beyond the reach of the senses, and less than the roots of the matter will never satisfy the scientific mind."[12] Judging from the writings of Alfred Mayer, certain American physicists in the 1870s were in general agreement with this mechanical viewpoint.

Sir William Thomson (1824–1907), made an extended visit to the United States in 1884, a decade after Tyndall's tour and eight years after his own earlier trip. He had first come, in 1876, to serve as a judge at the Centennial Exhibition in Philadelphia. During the late summer of 1884, he participated in the BAAS meeting in Montreal before traveling again to Philadelphia. There he served as vice-president of the Electrical Congress and presented a paper on the Hall effect at the concurrent meeting of the AAAS. A few weeks later, under the auspices of the Franklin Institute in Philadelphia, he also gave a popular talk on the well-worn topic, "The Wave Theory of Light." Finally, during the first few weeks of October, he went on to Johns Hopkins to deliver his "Baltimore Lectures" on "Molecular Dynamics and the Wave Theory of Light." Presi-

[73]

NOTES OF LECTURES

ON

Molecular Dynamics

and

The WAVE THEORY OF LIGHT

—

Delivered at the Johns Hopkins University Baltimore.

BY

SIR WILLIAM Thomson KELVIN

Professor in the University of Glasgow.

—

STENOGRAPHICALLY REPORTED BY

A. S. HATHAWAY,

Lately Fellow in Mathematics of the Johns Hopkins University.

—

1884.
Copy-Right by the JOHNS HOPKINS UNIVERSITY
BALTIMORE, MD.

Cover of William Thomson's "Baltimore Lectures," lettered by hand and reproduced "papyrographically." (Rare Book Department, University of Wisconsin Library) dent Gilman had arranged for Thomson to lecture at Hopkins, promising him that "your influence in the promotion of science in this country . . . would be very strong, and would extend far beyond the immediate companies of your hearers, and far beyond the period of your visit." His audience in Baltimore included Henry Rowland, Thomas Mendenhall, Henry Crew, Edward Morley, and Albert Michelson.[13]

It is particularly informative to analyze Thomson's "Baltimore Lectures" because they later gained prominence as the classic statement of his mechanical credo. It is also instructive to compare both of his 1884 American presentations on light with two presentations that we previously discussed: the 1884 speeches of Henry Rowland and John Trowbridge before the Electrical Congress and the AAAS. Fortunately, the texts of Thomson's presentations have survived: he retained and later published a copy of his popular talk in Philadephia, and a Johns Hopkins student with stenographic talents produced a longhand, verbatim transcript of the more technical Baltimore course.[14]

Thomson's goal in his two-and-one-half-week course at Johns Hopkins was to highlight the "outstanding difficulties" that still remained in treating the wave theory of light as part of molecular dynamics. With the kinetic theory of gases already "reduced to molecular dynamics," he felt that the wave theory of light was left as "the most important branch of physics which at present makes demands upon molecular dynamics." In his lectures, he focused on a group of basic problems concerning the dynamics of light: how to account for the detailed properties of dispersion, refraction, and reflection, as well as how to explain the specific dynamical qualities of the ether. Thomson did not doubt the existence of the ether; he was merely unsure of its exact dynamic makeup. Thus in his opening lecture at Hopkins, he insisted that "we must not listen to any suggestion that we must look upon the luminiferous ether as an ideal way of putting the thing. A real matter between us and the remotest stars I believe there is, and that light consists of real motion of the matter."[15] One member of Thomson's Baltimore audience, Henry Crew, similarly recorded in his diary that Thomson was "very confident in his assertions as to the reality of a luminiferous ether." In his Philadelphia talk—which somewhat anticipated the opening Hopkins lecture—Thomson similarly em-

[75]

phasized: "One thing we are sure of, and that is the reality and substantiality of the luminiferous ether."[16]

Thomson, while confident of the ether's existence, expressed uncertainty about its detailed elastic properties. At Hopkins, he carefully qualified as follows: "The luminiferous ether we must imagine to be a substance which so far as luminiferous vibrations are concerned moves as if it were an elastic solid. I do not say it is an elastic solid. That it moves as if it were an elastic solid . . . is the fundamental assumption of the wave theory of light." Thomson acknowledged that this assumption of an elastic-solid ether was at odds with the requirement that the ether be pliable enough to permit the free passage of celestial bodies through space. Nevertheless, he found strong support for the assumption—and for a reconciliation of the opposing ethereal properties of rigidity and penetrability—in an analogy between the ether and "Scottish shoemaker's wax." Such wax was both brittle enough to vibrate and fluid enough to yield to bodies passing through it.[17] In his Philadelphia talk, he similarly stressed the speculative nature of specific views on the ether's elasticity; but again he felt that "the illustration of a shoemaker's wax" supported the view of the ether as an elastic solid. In his closing sentence at Philadelphia, he thus stated: "When we can have actually before us a thing elastic like jelly and yielding like pitch, surely we have a large and solid ground for our faith in the speculative hypothesis of an elastic luminiferous ether, which constitutes the wave theory of light."[18]

Thomson's dependence on a commonplace, concrete "illustration" to justify his assumption of an elastic ether points to the feature for which the Baltimore lectures are best known today: his insistence on mechanical models in explicating the molecular dynamics of light. In his final lecture, he said: "I never satisfy myself until I can make a mechanical model of a thing. If I can make a mechanical model I can understand it. As long as I cannot make a mechanical model all the way through I cannot understand. . . ." In his eleventh lecture, he likewise stated: "My object is to show how to make a mechanical model which shall fulfill the conditions required in the physical phenomena that we are considering, whatever they may be. . . . It seems to be that the test of 'Do we or not understand a particular subject in physics?' is, 'Can we make a mechanical model of it?'" Thomson, however, added an impor-

[76]

tant qualification. Such "understanding" in terms of "mechanical models" did not, in the present state of scientific knowledge, have ontological import. Thus he actually began the preceding passage from his eleventh lecture with a cautionary note; although he "could not be satisfied" without a mechanical model, he stressed that a model "is not to be accepted as true in nature." Throughout his lectures he was careful to label each atomo-mechanical device a "rude mechanical model" or a "crude mechanical explanation." However cautious Thomson was with current models, he did view each of them as "a step toward a definite mechanical theory" that would be complete and enduring. Differentiating between illustrative models and true theories, he retained the goal of an eventual mechanical theory.[19]

Thomson's insistence on detailed mechanical models made him uncomfortable with the trend in electromagnetism which led away from concrete models toward general dynamical principles. He was particularly uncomfortable with Maxwell's effort, beginning in the mid-1860s, to formulate a purely electromagnetic theory of light.[20] Although he had "an immense admiration" for Maxwell's early, more traditional insights, such as his "mechanical model of electro-magnetic induction," Thomson objected to his later, more abstract outlook on light. "I want to understand light as well as I can," he wrote, "without introducing things that we understand even less of. That is why I take plain dynamics. I can get a model in plain dynamics, I cannot in electro-magnetics." Similarly, in his opening lecture, he stated: "It seems to me that it is rather a backward step from an absolutely definite motion that is put before us by Fresnel and his followers to take up the so-called Electromagnetic theory of light in the way it has been taken up by several writers of late."[21] Incidentally, it was in the context of this latter passage that Thomson exclaimed, "If I knew what the magnetic theory of light is, I might be able to think of it in relation to the fundamental principles of the wave theory of light." This did not mean, as certain individuals later implied, that Thomson at that moment was ignorant of Maxwell's electromagnetic theory of light; he was merely disdainful of the theory since it lacked detailed mechanical underpinning.[22]

How did physicists react to Thomson's Baltimore lectures? The most dramatic reaction came much later, in 1906, when the French

[77]

physicist and historian Pierre Duhem took Thomson to task for his distinctively English reliance on mechanical models. Thomson, in his Baltimore lectures, typified English physicists who, having "ample and weak" minds, were limited in their thinking to "concrete, material, visible, and tangible things." This was in contrast to French physicists who were strong-minded and who excelled in abstract thought. Duhem particularly criticized Thomson and his English colleagues for not securely anchoring their mechanical concepts to metaphysics, philosophy, and logic. Desiring from science "an explanation acceptable to reason," Duhem was disconcerted by "the very provisional character" of Thomson's various models in the Baltimore lectures.[23] Ironically, whereas Stallo had damned the atomo-mechanical viewpoint because of its metaphysical overtones, Duhem denounced the same viewpoint, particularly Thomson's, because it was metaphysically vacuous—a mere psychological habit characteristic of Englishmen.

But how did physicists attending the Baltimore lectures react in October 1884 to Thomson and his views on molecular dynamics, an elastic ether, and mechanical models? In a letter which Thomson's wife sent home to England, she reported that the lectures "are going on capitally, and he has a most eager, interested audience of about 60 or 70. . . ." Lord Rayleigh, after staying for only the first half of the course, wrote to his mother, "Our week at Baltimore was a success. Thomson had collected half the physicists in America, so that we had almost daily discussions." Rayleigh added, however, that the "lectures were quite in the usual Thomsonian style, a sort of thinking out loud in an enthusiastic incoherent manner." Such enthusiastic incoherence contributed to the harsh evaluation that Henry Crew, doctoral student at Hopkins, reported in his diary: "Sir Wm. Thomson finished his course of 19 lectures on Molecular dynamics this eve: it was a grand failure—so say all the professors who understood his math: he did not put things clearly." Henry Rowland was probably among the mathematically qualified professors who branded the course "a grand failure." His concept of light contrasted sharply with Thomson's. A month earlier at the Philadelphia Electrical Conference, Rowland had endorsed Maxwell's electromagnetic theory of light and suggested that it was reconcilable with conventional mechanical principles. Reportedly, he was bored by Thomson's musings and slept through the lectures.[24]

[78]

While certain Americans attending the Baltimore lectures were dissatisfied with Thomson's traditional message, the message itself had been neither myopic nor naive. Indeed, Thomson had frankly presented the problems confronting molecular dynamics and, to Duhem's later dismay, had emphasized that present mechanical models, while essential, were merely illustrative. American misgivings about even Thomson's self-critical and circumspect mechanical outlook suggest that by the mid-1880s, physicists were not as uniformly and deeply committed to the atomo-mechanical viewpoint as Stallo contended. We can reinforce this conclusion by now looking at American physical scientists who more sharply departed from conventional doctrines. It will become clear that in the decades around 1880 there was a distinct strain of nonmechanical thought.

QUALIFICATIONS, ALTERNATIVES,
AND DEVIATIONS

Secretary of the Smithsonian Institution, Samuel Langley. (Smithsonian Institution Photo No. A53209)

8

NEWCOMB'S OPERATIONAL OUTLOOK

Nonmechanical perspectives existed alongside the range of mechanical outlooks, but were less prevalent. Such minority views were not necessarily in open opposition to established atomomechanical positions; nevertheless, they were potentially corrosive of established viewpoints. At the same time, these nonmechanical views were not necessarily tied to fully developed research programs or to formal ideologies. By about 1900, however, they would evolve into distinct forces counterpoised against traditional mechanical physics. For these reasons, the nonmechanical orientations of American physical scientists in the decades around 1880 are most accurately described as qualifications, deviations, or nascent alternatives. Three such views were particularly prominent: Simon Newcomb's operationalism; J. Willard Gibbs's qualified phenomenalism; and Samuel Langley's and Francis Nipher's scientific skepticism. Incidentally, although all of these men were interested in physics, three of them were also deeply involved in other sciences—Newcomb and Langley in astronomy, Gibbs in chemistry, and Newcomb and Gibbs in mathematics. It is likely that these multidisciplinary involvements which ranged beyond the confines of mainstream physics correlated with the broader scientific perspectives of these three men.

Simon Newcomb (1835–1909) was widely recognized as a leading astronomer, mathematician, physicist, and political economist. Indeed, by the late 1870s he was one of the half-dozen or so most eminent and influential scientists in America, respected as both an internationally involved researcher and an articulate spokesman for the American scientific community. In 1877, he became superintendent of the American Nautical Almanac Office as well as president of the AAAS. At the time of his death in 1909, his list of professional achievements and awards was so extensive that even President Taft felt obliged to attend his elaborate state funeral.[1]

His career was launched in part by Joseph Henry, who at the same time was helping Alfred Mayer get started in science. With

Simon Newcomb, late 1870s. (Manuscript Division, Library of Congress)

Henry's guidance, Newcomb became, at the age of twenty-two, a computational assistant at the Nautical Almanac Office in Cambridge, Massachusetts. During these Cambridge years, he also studied mathematics, celestial mechanics, and physics under Benjamin Peirce at Harvard's Lawrence Scientific School, and obtained a Bachelor of Science degree in 1858. His interest in physics was especially pronounced during this period. At one point, he drafted a textbook of college physics, and earlier, he unsuccessfully sought a professorship in physics at Washington University in St. Louis. The allure of astronomy finally prevailed. In 1861, he joined the staff at the U.S. Naval Observatory in Washington and eventually rose to the position of head of the Nautical Almanac Office. While at these two governmental agencies, his reputation in mathematical astronomy was established by the undertaking of an exhaustive study of the motions and positions of the moon, planets, sun, and principal stars. He also maintained his interest in physics; around 1880, in close communication with young Albert Michelson, he redetermined the speed of light.[2]

Newcomb had a number of experiences in his younger years that led him toward an operational viewpoint—a viewpoint which held that concepts have unambiguous meaning only when defined in terms of actual experiences or measurements. Very important were his early exposures to the philosophies and scientific perspectives of Auguste Comte, Charles Darwin, and John Stuart Mill.[3] While still in Cambridge, he delved into Comte's positivism, Darwin's recent evolutionary theory with its phenomenalistic packaging, and Mill's empiricism. Mill, whom Newcomb actually met in 1870, was his favorite writer on the nature of scientific method and inquiry. Also important in his formulation of an operational viewpoint was his close relationship with two innovative American thinkers— Charles Peirce (1839-1914) and Peirce's mentor, Chauncey Wright (1830-1875), those much acclaimed fathers of American pragmatism. Wright and Newcomb were fellow "computers" during the late 1850s in the Cambridge office of the Nautical Almanac. Recalling this association, Newcomb wrote that "philosophical questions were our daily subjects of discussion." Charles Peirce and Newcomb were, in 1870, participants in a solar eclipse expedition to the Mediterranean. Later, in the early 1880s, both became

[85]

members of the Johns Hopkins faculty. At various times in their younger years, Newcomb and these two men had shared both the formal tutelage and the personal hospitality of Benjamin Peirce, Charles's father and prominent professor at Harvard and the Lawrence Scientific School.[4]

It was not until the 1870s, when he was around forty, that Newcomb regularly begain to write about the philosophical, theological, and social implications of science. In one of the earliest of these public musings on "the spirit of modern science," written in 1874, Newcomb displayed general phenomenalistic leanings. He declared that the "first proposition" of modern science is that its "methods and objects . . . are distinguished by their purely practical character, using the word 'practical' in its best sense. Indeed, the most marked characteristic of the science of the present day . . . is its entire rejection of all speculation on propositions which do not admit of being brought to the test of experience." Taking as an example the prediction of an eclipse using the law of gravitation, he further asserted: "This prediction is complete with respect to the phenomena and to everything connected with it which can influence the material interest of mankind, yet it is entirely independent of the question, What causes the moon to gravitate toward the earth and sun?"[5]

This attention to "practical" consequences and concrete phenomena as well as the rejection of metaphysical speculation represented, by nineteenth-century standards, an astute attitude, but it was an attitude shared by a growing number of thinkers following in the paths of Comte, Darwin, Mill, and others. Moreover, although Newcomb's phenomenalism provided a foundation conducive to a specifically operational superstructure, he did not explicitly stress definitions of individual concepts in terms of sensory experiences or concrete measurements—the stress on language and meaning characteristic of an operational outlook. This particular emphasis was not to appear in Newcomb's writings until late 1878. Significantly, this occurred immediately after the publication of Charles Peirce's only early pragmatic essay and soon after the republication of Chauncey Wright's various essays having operational leanings.

Chauncey Wright died in Cambridge at the age of forty-five, a man unknown to most of his contemporaries but highly respected as a lay philosopher and scientist by an informal circle of colleagues and protégés which included Charles Peirce and William James. Just before his death in 1875, Wright published one of his clearest operational statements. He presented the statement, however, in a muted manner without special emphasis as part of an extended book review titled "Speculative Dynamics." In reviewing a recent book on the "mechanics of the universe," Wright sharply criticized the author, a nonscientist, for employing scientific terms to serve philosophically speculative ends. Regarding the physicists' term "force," Wright insisted: "All its uses in mathematical language . . . refer to precise, unambiguous definitions in the measures of the phenomena of motion, and do not refer to any other substantive or noumenal existence than the universal inductive fact that the phenomena of all actual movements in nature can be clearly, and definitely, or intelligently analyzed into phenomena, and conditions of phenomena, of which these terms denote the measures." Similarly, he explained that physicists had in mind particular "sensible properties" and "sensible measures" when they referred to "stores of energy" (i.e., potential energy). Incidentally, Wright had already revealed such operational leanings as early as 1865 in his lengthy article "The Philosophy of Herbert Spencer." This was a scholarly attack on the then popular Spencer for misappropriating and obfuscating the clear, experiential concepts of physics and biology.[6]

Charles Peirce expressed views similar to those of Wright. Although Peirce worked during his middle years as a gravitational specialist in the United States Coast Survey, his main interest until his death in 1914 was philosophy. His pragmatic outlook, while better focused than Wright's, appeared originally as just one briefly used philosophical weapon in what was his growing, maturing, and ever-changing arsenal. Following William James's lead, most scholars seeking the American roots of pragmatism point to a single paper written by Peirce—"How to Make Our Ideas Clear"—a paper which he published in 1878 but formulated during the early 1870s while meeting in the informal "Metaphysical Club" with Wright, James, and other Cambridge thinkers. Embedded within this 1878 paper were two sentences that later became known as Peirce's "pragmatic maxim": "Consider what effects, that might conceivably have practical bearings, we conceive the object of our

[87]

conception to have. Then, our conception of these effects is the whole of our conception of the object." In illustrating this rule, Peirce specifically echoed Wright in rejecting metaphysical interpretations of scientific concepts—interpretations that were not founded on "sensible perception" or "conceived sensible effects." Thus, regarding the term "force" Peirce wrote: "The idea which the word force excites in our minds has no other function than to affect our actions, and these actions can have no reference to force otherwise than through its effects. Consequently, if we know what the effects of force are, we are acquainted with every fact which is implied in saying that a force exists, and there is nothing more to know."[7]

Peirce's paper appeared in the January 1878 issue of *Popular Science Monthly,* the second of a series of articles by him titled "Illustrations of the Logic of Science." During the previous year, 1877, at the instigation of Chauncey Wright's small circle of admirers, Wright's main writings—including "Speculative Dynamics" and "The Philosophy of Herbert Spencer"—appeared in a posthumous volume called *Philosophical Discussions.* In his autobiography, Newcomb later praised Wright's papers in this 1877 collection, saying "their style is clear-cut and faultless in logical form." Similarly, as early as 1876, Newcomb had publicly complimented Wright as one of the few American scientists having "philosophical comprehension, scientific accuracy, and clearness of thought."[8] Given the publication dates of these writings by Peirce and Wright, and given Newcomb's unfailing admiration for Wright and close acquaintanceship with Peirce, it seems more than coincidence that Newcomb's operational emphasis first emerged in his public utterances late in 1878. The parallels between the writings of Newcomb and his two colleagues, particularly Wright, strengthen this supposition of intellectual ties.

In August of 1878, as retiring president of the AAAS, Newcomb gave a major address titled, "The Course of Nature." Various American journals, including *Popular Science Monthly,* and at least one British journal reprinted the text of the speech.[9] Anticipating this broad audience, Newcomb likely approached the speech with deliberation. What were his considered views on science and the "course of nature"? After some introductory niceties, Newcomb proceeded to his "keynote," beginning with a restatement of his general, phenomenalistic "proposition" first presented in 1874. "This proposition is, that science concerns itself only with phenomena and the relations which connect them, and does not take account of any questions which do not in some way admit of being brought to the test of observation. . . ." In the next few sentences, Newcomb added a new twist to this familiar proposition, molding and shaping it into a rudimentary operational maxim: "To speak with a little more precision, we may say that, as science only deals with phenomena and the laws which connect them, so all the terms which it uses have exact literal meanings, and refer only to things which admit of being perceived by the senses, or, at least, of being conceived as thus perceptible." This precise statement closely parallels Peirce's pragmatic maxim published a half-year earlier and his accompanying discussion of the meanings of words in terms of "sensible perception" and "conceived sensible effects." It also parallels Wright's previous emphasis on "precise, unambiguous definitions" in terms of "measures of the phenomena" and "sensible properties."

Newcomb ended this introductory "keynote" with an observation recalling the recurrent laments expressed by Peirce and Wright regarding the misuse and blurring of scientific concepts by metaphysically minded philosophers: "The use of plain language appears to be an actual source of difficulty with some in trying to understand the philosophy of science. Long habit in the use of figurative language in which ideas not readily comprehensible are symbolized by common terms leads one to look for hidden meanings in all philosophic discourse, and to see difficulties in terms which, to a scientific thinker, are as plain and matter-of-fact as an order for breakfast to an hotel-waiter."[10] Newcomb, with this address before the AAAS, adopted a viewpoint akin to the one held by Wright and Peirce. He also moved beyond Wright and Peirce, however, by refining their type of operational outlook, self-consciously making it the essence of his attitude toward science.

Two years after the AAAS speech, Newcomb in another major address revealed his continued emphasis on precise language and meaning. As president, he spoke to the Philosophical Society of Washington in late 1880 on "The Relation of Scientific Method to Social Progress." Again the address was widely disseminated: it was published as a separate pamphlet, reprinted in two journals,

[89]

and then reissued in 1906 in a collection of his essays and addresses.[11] Newcomb here argued for the application of scientific methods to current political and social as well as philosophical problems. Specifically, he proposed an operational mode of attack, advocating a "community of understanding" regarding social issues which can be brought about by disciplining persons to use "proper definitions" of relevant terms:

> The scientific discipline to which I ask mainly to call your attention consists in training the scholar to the scientific use of language. Although whole volumes may be written on the logic of science there is one general feature of its method which is of fundamental significance. It is that every term which it uses and every proposition which it enunciates has a precise meaning which can be made evident by proper definitions.

What in particular were "proper definitions"? Newcomb answered a few pages later with an example; he considered how we might communicate with "a person of well-developed intellect, but unacquainted with a single language or word that we use." "Every term which we make known to him," Newcomb concluded, "must depend ultimately upon terms the meaning of which he has learned from their connections with special objects of sense." Apparently guarding against the charge of being a naive sensationist or extreme empiricist, Newcomb went on to add that "the mind, as well as the external object, may be a factor in determining the ideas which the words are intended to express." "Whatever theory," he nevertheless insisted, "we may adopt of the relative part played by the knowing subject, and the external object in the acquirement of knowledge, it remains none the less true that no knowledge of the meaning of a word can be acquired except through the senses, and that the meaning is, therefore, limited by the senses."[12]

Although Newcomb, with his sustained stress on the operational aspects of science, continued in this 1880 speech to differ from Wright and Peirce, he still frequently repeated their earlier opinions, particularly those of Wright. (Wright, in turn, frequently restated the ideas of Mill.) Newcomb's discussion of the word "cause" as a purely neutral, descriptive term, for example, closely reflected a similar discussion in Wright's most famous essay, "Evolution of Self-Consciousness" (1873, reprinted 1877), which was publicly praised by Darwin.[13] An even more obvious example of borrowing from Wright involves an episode in the history of science that Wright originally characterized in his "Speculative Dynamics" (1875, reprinted 1877). Newcomb essentially paraphrased Wright's account of the controversy circa 1700 over the meaning of the phrase "force of a moving body" wherein natural philosophers were asking if "force" was proportional to velocity or to velocity squared. Like Wright, Newcomb pointed to this dispute to illustrate that disagreements frequently arise because the opposing parties do not underpin their seemingly contrary terminology with precise definitions relatable to concrete measurements.[14]

That Newcomb did sustain an explicit operational outlook during mid-career is further and finally illustrated in his scathing review of Stallo's *Concepts and Theories of Modern Physics*—a review which, as we saw in chapter 3, Stallo rebutted in *Popular Science Monthly.* This extended book review, published during 1882 in the New York-based *International Review,* contained Newcomb's most succinct operational statement. Beginning with specific examples of Stallo's "total misconception of the ideas and methods of modern science," Newcomb wrote:

> The word *mass,* for instance, as commonly used in physics, is an abstract noun like *length;* but he [Stallo] uses it as a concrete term, and in nearly the same sense as we commonly use the word matter. He speaks of the conservation of both mass and motion in a way which shows entire unconsciousness of the fact that this expression has no meaning at all. To give it a meaning we must first define the method in which mass and motion are to be measured, and then, in so many ways as we choose to make this measurement, just so many meanings may the expression have. A bar of metal, for instance, may be measured by its length, its breadth, its solid contents, or its weight. A pile of such bars may be measured by putting them end to end or piling in various ways, and measuring the length of the pile in as many ways as we choose. So, in measuring the motion of a system of bodies, we may adopt almost an infinity of different ways which will give different results. The very first necessity of any exact scientific proposition is a definition, without ambiguity, of a precise method in which every quantitive measure brought in shall be understood. The conclusions are then valid, assuming that particular method of measurement, but they are not valid on any other method.

Newcomb proceeded to use this operational dictum as a basis for his detailed critique of Stallo's book.[15] In doing this, he again echoed Chauncey Wright's work, particularly his 1875 book review, "Speculative Dynamics," that contained the clear operational statement we examined earlier. That is, Newcomb's criticism of Stallo's analysis of physics was reminiscent of Wright's appraisal of an earlier nonscientist's philosophical interpretation of physics. Newcomb even gave his review a parallel title, calling it "Speculative Science."[16]

To understand more fully Newcomb's clash with Stallo, we must notice that during the middle of his career Newcomb aimed his operational salvos mainly at theologians and philosophers who were overstepping their legitimate domains of inquiry: he used operational notions expressly to attack metaphysical or teleological explanations of the physical universe. Occasionally, he also fired an operational barrage at public officials or leaders of society who were hindering social progress with their "ambiguous" interpretations of issues. He implicitly assumed that the operational viewpoint was the controlling viewpoint of qualified scientists; thus he pictured himself as bringing modern scientific attitudes to bear on the confused terminology, and hence confused ideas, of theologians, philosophers, and other uninitiated thinkers. Generally speaking, Wright and Peirce shared this scientific self-righteousness and missionary spirit, a self-righteousness and spirit generated largely through the ongoing Darwinian controversy. It was over the issues of evolution, Newcomb wrote in 1879, that the dispute between scientists and humanists was currently "raging with most bitterness."[17]

Newcomb's general perception that scientists were free of metaphysical prejudices translated into his specific trust in the prevailing atomo-mechanical framework of physical science. To put it another way, Newcomb attacked the outlooks of nonscientists while simultaneously and implicitly defending the views of nineteenth-century physicists. Considering modern science inherently above dispute, he concurred with Alfred Mayer that physical phenomena were ultimately understandable in terms of underlying atoms and a material ether that obeyed the laws of classical mechanics. As Newcomb asserted in 1880, "The hypothesis upon which all investigation proceeds . . . is that every attribute of a body can be explained in a scientific sense by its internal structure and by mechanical forces at play among its molecules." Even though he accepted this traditional research program with its "ultimate atoms" and "ethereal medium," Newcomb did stress its hypothetical and provisional nature. That is, he perceived himself to be free of epistemological and ontological delusions regarding the attainment of absolute knowledge of physical reality. To say that a proposition is scientifically true, he wrote in 1878, does not mean "that it bears any recognized seal of truth" but only "that the evidence in favor of it entirely preponderates over all that can be brought to bear against it."[18] Along with Newcomb, Charles Peirce also accepted atomomechanical physics during the period prior to the mid-1880s. In the same series of articles that contained his famous pragmatic maxim, he wrote that it was "legitimate" for scientists to explain cosmological events in terms of "the fortuitous concourse of atoms"; after all, he added, "matter is supposed to be composed of molecules which obey the laws of mechanics. . . ." Peirce even held that such scientific insights were not merely provisional, but actual steps toward "truth and reality." Having become during his student years "thoroughly grounded . . . in all that was then known of physics and chemistry,"[19] Peirce, like Newcomb, had internalized many of the dominant precepts and attitudes of mechanical physics.

Realizing that Newcomb coupled an operationalist's distaste for the metaphysical speculations of nonscientists with an orthodox practitioner's commitment to an avowedly hypothetical atomomechanical research program, we can begin to understand Newcomb's otherwise bewildering attack on John Stallo's *Concepts and Theories of Modern Physics*. The attack was bewildering because on the level of general principle Stallo and Newcomb had similar scientific ideologies. Although Stallo did not mirror Newcomb's particular emphasis on sensate definitions, he did have strong phenomenalistic and antimetaphysical leanings, as both Ernst Mach and Percy Bridgman were to recognize in later years.[20] Thus, both Stallo and Newcomb would have opposed, to use Stallo's phraseology, "latent metaphysical elements" of thought and "ontological prepossessions"; both would have wanted "to foster and not to repress the spirit of experimental investigation, and to accredit instead of discrediting the great endeavor of scientific research to gain a sure foothold on solid empirical ground. . . ."[21] Why then did Newcomb so vigorously castigate a kindred spirit?

[93]

Why did he dismiss Stallo as another nonscientific "paradoxer," "charlatan," or "pretender" and sarcastically see the only purpose of Stallo's book as "misleading its readers" or "showing the possible aberrations of an evidently learned and able author"?[22]

In brief, Stallo precipitated Newcomb's attack by directing his antimetaphysical critique at the very foundations of the prevailing atomo-mechanical physics. Perhaps even more vexing to Newcomb, this heresy was coming from an outsider, a man with a "total misconception of the ideas and methods of modern science." To the xenophobic Newcomb, Stallo was just another nonscientific, philosophical speculator naively and ineffectually tilting with the solid and dependable windmills of modern physics.

Stallo, of course, was unequivocally opposed to the atomo-mechanical theory. In his early chapters, for instance, he had exhibited for analysis and criticism four ontologically laden propositions that supposedly constituted the conceptual foundation of current mechanical theory. Newcomb objected to associating these simplistic propositions with what he perceived to be the seasoned mechanical theory of the present day. To demonstrate the untenability of the four propositions—and thus to reveal that Stallo's critique was directed against fictional propositions—Newcomb invoked his operational dictum. For example, regarding the proposition that "the elementary units of mass, being simple, are in all respects equal," Newcomb responded: "This may well be true; but it is a proposition to which science would admit of no proof except from the facts. An attempt to prove it a priori as the author does, belongs entirely to a past age of thought." Similarly, Newcomb objected to characterizing "ultimate atoms" with phrases like "absolutely hard and inelastic" or "absolutely inert, and therefore passive"; such phrases had "no correct meaning" in that they were neither derived from "qualities of sensible masses" nor from "effects which follow when matter is placed under certain conditions." These admonitions against attributing the quality of hardness to atoms, by the way, paralleled Peirce's 1878 explication of the pragmatic meaning of hardness.[23]

In his lengthy rebuttal in *Popular Science Monthly,* Stallo zeroed in on what he considered to be a gross misinterpretation made by Newcomb. Adding to an already confused dialogue, Stallo understood Newcomb to be attacking the *Concepts and Theories of*

Modern Physics for its *endorsement* of the four propositions and its *defense* of a naively metaphysical interpretation of atomo-mechanical theory. To Stallo, his own sustained critique of the mechanical theory was "insusceptible of misapprehension even by the most hebetated intellect." He was able to explain what he took to be Newcomb's misinterpretation only by supposing that Newcomb decided "no doubt before writing his article" that Stallo was a "dogmatic defender" of the atomo-mechanical theory and, particularly, the four ontological propositions. Stallo went on to point out, ironically, that his case against current-day physics was enhanced by Newcomb's criticism of the propositions: "In this way he levels his thrusts at the most eminent physicists and mathematicians of the day, laboring always under the hallucination that he is striking at me."[24]

Although Newcomb and Stallo actually had a similar distaste for metaphysical interpretations of physics, Newcomb would never have agreed with Stallo that such interpretations were truly held by "eminent physicists and mathematicians." Thus Stallo's critique of the kinetic theory of gases especially ruffled Newcomb. "The scientific reader," Newcomb wrote, "will probably find most amusement in the author's attack on the kinetic theory of gases"; after all, "there is no theory of modern physics, the processes supposed by which are invisible to direct vision, which is more thoroughly established than this." Responding to Stallo's specific objection that the kinetic theory was incompatible with the fundamental proposition that molecules are inelastic, Newcomb sidestepped this dilemma by adding: "No elasticity is assigned the molecules in the kinetic theory, but only an insuperable repulsive force which causes the molecules to repel each other when they are brought sufficiently near together."[25]

In his rebuttal, Stallo pounced upon this defense of the kinetic theory of gases. On the specific issue of elasticity, he argued that prominent physicists, William Thomson for one, were denying the fundamental postulate of inelasticity by assuming that molecules in the kinetic theory were elastic. Moreover, he sarcastically derided Newcomb for attempting to replace "elasticity" with the even more nebulous and empirically untenable concept of an "insuperable repulsive force" between molecules. Finally, in closing his appraisal of Newcomb's view on the kinetic theory, Stallo drew a

[95]

broader lesson: "It is not a little instructive to note the character of sacredness ascribed by persons of Professor Newcomb's frame of mind to dominant physical theories, and the violence with which they repel every attempt to point out their defects."[26] While Newcome viewed Stallo as a carping charlatan, Stallo thus viewed Newcomb as a near-sighted dogmatist.

In actuality, as we have seen, both Stallo and Newcomb were adamant in their advocacy of nonmetaphysical, empirical thought; but Stallo alone directed his critique at the conceptual core of modern science. Newcomb, with his operational liturgy but atomo-mechanical faith, appears to be a man of curious ambivalences. Committed to an operational perspective when admonishing non-scientists, he seems reluctant to turn that same perspective against the concepts used by his own colleagues. Of course, Newcomb himself would disagree that he was ambivalent; he perceived that mechanical physics unquestionably had adequate empirical foundations. Nevertheless, as we will see, American physical scientists by the end of the century were increasingly moving away from Newcomb's self-confident, outward-looking operationalism toward a self-critical, inward view. Contrary to his expectations, Newcomb in the period around 1880 helped enunciate an outlook potentially corrosive of his own atomo-mechanical physics.

9

GIBBS'S PHENOMENALISTIC LEANINGS

Newcomb elaborated clear operational precepts but seldom applied them to the everyday workings of physics. J. Willard Gibbs, on the other hand, practiced a phenomenalistic style of physics but avoided methodological orations. Gibbs's leanings toward phenomenalism appeared in his research papers on thermodynamics, written largely in the mid-1870s. This tendency was also evident in his technical studies on the electromagnetic theory of light, done in the 1880s, and in his work on statistical mechanics, done during the 1880s and 1890s. His phenomenalistic leanings, however, were just that—leanings. There was a discernible atomo-mechanical current throughout his work. For the most part, he meticulously constructed his thermodynamics from broad, descriptive principles related to observable, sensory quantities, but occasionally and without hesitation, he fell back on corpuscular interpretations. Moreover, he followed his thermodynamics with studies on statistical mechanics. Even in this branch of atomo-mechanical science, however, he adopted a stance that was empirically cautious as well as logically and mathematically rigorous. Finally, he rejected earlier theories of light that postulated various elastic ethers and specific atomic mechanisms; instead, he championed the refined, mathematically strict, Maxwellian form of analytic dynamics that rested firmly on experiment. Thermodynamics, statistical mechanics, and electromagnetism would, by about 1900, give rise to the three main substantive threats to nineteenth-century atomo-mechanical views; Gibbs, however, was still far from a nonmechanical delineation of these three fields. As developed by Gibbs, thermodynamics, statistical mechanics, and electromagnetism furnished nascent alternatives to traditional mechanical physics.

The son of a Yale professor of sacred literature, J. Willard Gibbs (1839–1903) attended and later taught at his father's college.[1] Excelling in mathematics and Latin, he obtained his undergraduate degree in 1858 and then went on, in 1863, to earn a Ph.D. in engin-

The young J. Willard Gibbs. (Burndy Library, courtesy of AIP Niels Bohr Library)

eering—the first doctorate in engineering awarded in the United States. After tutoring at Yale for a few years, he journeyed abroad. Remaining in Europe for three years, he refined his knowledge of physics and mathematics by reading privately and by attending lectures at the universities of Paris, Berlin, and Heidelberg. Back in New Haven and at Yale, Gibbs, in 1871, became professor of mathematical physics, a post he held until his death in 1903.

Given Gibbs's publication of his early papers in an obscure Connecticut journal and his initially unsalaried professorship at Yale, we might suspect that this reticent and unassuming scholar was unnoticed and unappreciated by his American colleagues. Maxwell's immediate enthusiasm for Gibbs's thermodynamics in the mid-1870s, along with subsequent endorsements by Europeans such as Wilhelm Ostwald, might lead us to conclude that Gibbs found his deserved recognition only among erudite foreigners, to whom he regularly sent his papers. The truth was that he was an integral member, if not an acknowledged leader, in the American scientific community by about 1880, only a few years after completing his main thermodynamic writings. His contribution to thermodynamics led Henry Rowland in 1879 to tell him that Americans were "proud to have at least one in the country who can uphold its honor" in the realm of mathematical physics. This pride translated a few months later into an invitation to Gibbs from President Daniel Gilman to give a guest lecture at Johns Hopkins, and then in 1880, an invitation to join the Hopkins faculty. In recommending this appointment to Hopkins, which Gibbs eventually declined, Rowland proudly stressed both Maxwell's private and published praise of Gibbs.[2]

It was also in the years 1879 and 1880, when Gibbs was around forty, that his countrymen further honored him by electing him to both the National Academy of Sciences and the American Academy of Arts and Sciences. Moreover, this latter group awarded him its prestigious Rumford Medal in 1881; Gibbs, ever the recluse, had John Trowbridge accept the award in his name. By 1884, young American scientists such as Albert Michelson were corresponding regularly with Gibbs, seeking his advice. Even John Stallo requested a copy of a recent publication. Apparently Gibb's reclusive personality did not harm his scientific reputation; on the eve of the 1884 Philadelphia Electrical Conference, to which Gibbs had been invited as an official representative of the United States, president Rowland pleadingly wrote: "Can you not accept even if you cannot come as we wish the weight of your name." This unassuming but respected American scientist accepted and made the trip, one of his rare professional sojourns away from New Haven.[3]

Gibbs's phenomenalistic leanings were most clearly discernable in his thermodynamic theories. Following in the tradition of nineteenth-century scientists Sadi Carnot and Rudolf Clausius, Gibbs built his thermodynamics on a few broad principles directly relatable to observable quantities. He wrote three main papers each stressing basic mathematical relations between the macroscopic variables of volume, pressure, temperature, and, particularly, energy and entropy. The first two papers, published in 1873, dealt with two- and three-dimensional graphical or geometrical representations of thermodynamic properties. In his third and most famous paper—a lengthy monograph published in parts during 1876 and 1878, titled "On the Equilibrium of Heterogenous Substances"—Gibbs achieved his greatest degree of phenomenalistic simplicity and generality. He emphasized the all-embracing variables of energy and entropy, and also extended the domain of thermodynamics to include the equilibrium states of material systems that were not homogeneous (e.g., mixtures of chemicals). In an abstract of the paper in the *American Journal of Science,* Gibbs justified this approach by pointing out that it incorporated "those quantities which are most simple and most general in their definitions, and which appear most important in the general theory of such systems." Similarly, in accepting the Rumford Medal in 1881, he stressed that he had presented an analytic "process which seems more simple, and which lends itself more readily to the solution of problems, than the usual method, in which the several parts of a cyclic operation [involving exchanges of heat and work] are explicitly and separately considered."[4]

The most distinctive and innovative feature of Gibbs's approach to thermodynamics—indeed, the feature that gave the approach analytic power and phenomenalistic elegance—was the emphasis on entropy. As Martin Klein has pointed out,[5] during a period when physicists such as Maxwell and Tait were still misapplying and arguing about Rudolf Clausius's entropy interpretation of the second law of thermodynamics, Gibbs grasped its significance and elevated entropy to a position alongside energy in the hierarchy of thermodynamic concepts. Thus, in footnotes to both of his 1873 papers he carefully explained that he was using the term "entropy" as defined in 1865 by Clausius, and not as distorted by Tait and then Maxwell, who followed Tait's lead. He underscored this point in 1876 by opening his paper, "Equilibrium of Heterogeneous Substances," with Clausius's statement, in German, of the two laws of thermodynamics: "The energy of the world is constant. The entropy of the world tends toward a maximum." In an abstract of this paper, he similarly stressed the primacy of the principle that when a system's entropy has increased to a maximum, then the system is at equilibrium. "Although this principle has by no means escaped the attention of physicists," he added, "its importance does not appear to have been appreciated. Little has been done to develop the principle as a foundation for the general theory of thermodynamic equilibrium."[6] Gibbs developed the principle, and in so doing gained the respect of Maxwell and subsequently of other American and European physicists. These supporters included Wilhelm Ostwald, the physical chemist who, in 1892, translated Gibbs's three thermodynamic papers into German.

With his dedication to directly measurable variables, a few broad principles, lucid logic, and meticulous mathematics, Gibbs appeared to epitomize the scientific phenomenalist. Indeed, by the turn of the century, European "energeticists" Georg Helm and Wilhelm Ostwald claimed Gibbs as an ally in their campaign against atomism and mechanism.[7] In actuality, like Clausius before him, Gibbs unselfconsciously and unapologetically supplemented his "pure" thermodynamics with speculations on the corpuscular composition of substances. For example, toward the middle of his lengthy paper, "Equilibrium of Heterogeneous Substances," he interjected an atomo-mechanical analysis to unravel an unusually knotty problem. To explain why his macroscopic equations predicted an entropy increase upon mixing *identical* gases—an unreasonable prediction, later called the "Gibbs paradox"—he turned to a discussion of gaseous "particles," "atoms," and "molecules" and their "positions," "attractions," and "paths." In this shift from "sensible" to "molecular" properties of gases, moreover, he interpreted the second law of thermodynamics (that entropy always increases toward a maximum) as a statistical rather than

[101]

absolute law. Following the recent lead of Maxwell and Boltzmann, he concluded that "the impossibility of an uncompensated decrease of entropy seems to be reduced to improbability."[8] Not only was Gibbs willing to discuss in his thermodynamic papers the mechanics of atoms, but he espoused what was, for his time, an advanced statistical view of that mechanics; recall that six years later, in 1882, Stallo was still denouncing Maxwell's statistical outlook as irrational.

In an 1889 obituary of Clausius, Gibbs had occasion to recount the relative merits of the thermodynamic and molecular outlooks. Once again, he seemed appreciative of both outlooks, although he implied that thermodynamics was more fully developed than atomo-mechanical science. Thanks to Clausius, the foundations of thermodynamics "were secure, its definitions clear, and its boundaries distinct." But even though Clausius had initiated modern thermodynamics, his attention had been "less directed toward the development of the subject in extension, than toward the nature of the molecular phenomena of which the laws of thermodynamics are the sensible expression." Gibbs was enthusiastic about Clausius's overall contribution to "molecular science." Nevertheless, since certain of Clausius's propositions involved "quantities which escape direct measurement," Gibbs felt that they still required "for their complete and satisfactory demonstration a considerable development" of molecular science. After all, even the simplest department of that science, the kinetic theory of gases, was presently far from completion, although the theory did have "an extensive and well established body of doctrine." Gibbs was generally optimistic that molecular science would continue to advance. As he had hinted more than a decade earlier, future breakthroughs would come by following the "theory of probabilities" as proposed by Maxwell and Boltzmann—that is, by following statistical mechanics.[9]

Just as thermodynamics would in fact mature into a full-fledged rival to traditional atomo-mechanical outlooks, so too would statistical mechanics. Thermodynamics would represent a break from the past, while statistical mechanics would appear as an updating or recasting of traditional outlooks. Contradictory though it may sound, this new atomic mechanics would have a phenomenalistic flavor. Scientists would eschew detailed assumptions about the underlying nature of matter or the constitution of atoms in favor of a few broad, empirically cautious assumptions about statistical groupings of nondescript particles. As he had done in thermo-dynamics, Gibbs also pioneered in statistical mechanics.

He had already touched on probabilistic interpretations in his thermodynamic paper of 1876 and in his obituary of Clausius from 1889. In the intervening period, he prepared a more extended expo-sition, titled "On the Fundamental Formula of Statistical Mechan-ics with Application to Astronomy and Thermodynamics." He presented this at the 1884 meeting of the AAAS while attending the concurrent Electrical Conference in Philadelphia. Only the "Abstract" of this paper has survived, but from it we glimpse Gibbs's concern for phenomenalistic generality and simplicity. In deriving his statistical formula, he merely postulated "a great number of systems which consist of material points and are iden-tical in character, but different in configuration and velocities, and in which the forces are determined by the configurations alone." During the late 1880s and 1890s, he refined his views on statistical mechanics through various courses he taught at Yale. By 1892, he was able to write to Lord Rayleigh that he was "trying to get ready for publication something on thermodynamics from the a priori point of view, or rather on 'Statistical Mechanics.'"[10] As we will see in chapter 12, his thinking culminated in the influential text-book, *Elementary Principles in Statistical Mechanics,* published in 1902, just before his death.

Besides Gibbs's ground-breaking endeavors in thermodynamics and statistical mechanics, for which he is best known, he also pioneered in the electromagnetic theory of light. Again displaying phenomenalistic proclivities, he opposed the multitude of detailed atomistic and ethereal speculations in traditional mechanical theo-ries of light, especially as developed by William Thomson. Along with Henry Rowland, he favored the few, broad, empirically well-founded principles of Maxwell's electromagnetic theory of light. Around 1900, when generalized and extended to all physical phe-nomena, such an electromagnetic view would become a principal option to traditional atomo-mechanical physics.

Gibbs published five main papers on light: a group of three in 1882 and 1883, and another two in 1888 and 1889, all appearing in the *American Journal of Science.*[11] In the earlier ones, his goal was to demonstrate the efficacy of the electromagnetic theory in accounting for certain complex but basic optical phenomena: double refraction, dispersion of colors, and circular polarization. Believing that Maxwell had retarded the acceptance of the electromagnetic interpretation of light by initially connecting it to an atomo-mechanical "theory of electric action," Gibbs chose to build on the mature Maxwell's general dynamical principles, especially as presented in 1873 in his *Treatise on Electricity and Magnetism*. Thus, Gibbs dealt only with "plane waves of homogeneous light, regarded as oscillating electrical fluxes" that impinged on an optical body postulated to have "a very fine-grained structure . . . [of] dimensions very small in comparison with the wavelength." This uncomplicated view assured that "the average electrical displacement of such [fine-grained] parts of the body may be expressed as a function of time and the coordinates of position by the ordinary equations of wave-motion." Gibbs was especially careful to speak only of experimentally observable *average* displacements, as distinguished from "the *actual* electrical displacements, which are too complicated to follow in detail, and which in their minutiae elude experimental demonstrations."[12] Further, when speaking of his own ideas he avoided the word "ether." Working out the implications of his basic principles, he concluded that not only did they account for double refraction, dispersion, and polarization, but they did so better than the traditional elastic theory. "In no particular," he surmised, do the principles "conflict with the results of experiment, or require the aid of auxiliary and forced hypotheses to bring them into harmony therewith. . . . In this respect, the electromagnetic theory of light stands in marked contrast with that theory in which the properties of an elastic solid are attributed to the ether."[13]

In his final two papers, written in 1888 and 1889, Gibbs compared, point by point, the elastic-solid and electrical theories of light. He was motivated to write the 1888 paper by "the confusing multiplicity of the elastic theories" as well as by the recent availability of precise measurements on double refraction made by his former student, Charles Hastings. With "simplicity and generality," Gibbs highlighted the logical and empirical difficulties confronting the elastic theory but not the electrical.[14] A year later, he narrowed his analysis to a "remarkable" new proposal made by William Thomson that had "opened a new vista in the possibilities of the theory of an elastic ether." Admitting that Thomson's new theory of a "quasi-labile ether" provided the electrical theory with a "serious rival" in functionally accounting for optical phenomena, Gibbs nevertheless continued to opt for the electrical. "It may still be said for the electrical theory," he summarized, "that it is not obliged to invent hypotheses, but only to apply the laws furnished by the science of electricity, and that it is difficult to account for the coincidences between the electrical and optical properties of media, unless we regard the motions of light as electrical." The nonhypothetical status of "electrical motions," he added in a final footnote, had recently been reinforced by the experiments of Heinrich Hertz.[15]

Gibbs's phenomenalistic leanings were pronounced and readily apparent to careful readers. Seeking empirical, logical, and mathematical rigor, he anchored his thermodynamics to Clausius's two broad laws; his statistical mechanics and electromagnetic theory of light he moored to a few general dynamical principles. Very evident also were Gibbs's atomo-mechanical leanings. He supplemented his thermodynamics with atomistic interpretations and went on to develop a statistical mechanics. And his statistical mechanics and electromagnetism, although phenomenalistic in tone, were still within the traditional atomo-mechanical domain, being dynamical treatments of postulated particles. Regardless of this inherent ambivalence, Gibbs had demonstrated the phenomenalistic potential and analytic capability of three sciences which, in subsequent years, would evolve into distinct alternatives to nineteenth-century atomo-mechanical physics.

10

LANGLEY AND NIPHER: SKEPTICS

Astrophysicist Samuel Langley and physicist Francis Nipher, from the late 1870s through the early 1890s, forcefully affirmed the inherent fallibility of science. Using historical, philosophical, and psychological analyses, as Stallo did, they specifically questioned the popular, Victorian image of scientists directly progressing toward full knowledge of absolute natural laws. While physicists such as John Trowbridge were urging in occasional caveats a similar wariness toward claims of scientific omniscience, Langley and Nipher, in their various nontechnical writings, formulated more categorical and coherent statements of this skeptical attitude.

Such skepticism was a deviation from mainstream mechanical thought. To doubt science's infallibility was to doubt one of the basic ideological tenets traditionally associated with atomo-mechanical content. Recall, for example, that Alfred Mayer linked his confidence in the substantive content of mechanical physics to a faith that physics was making real progress in penetrating the "constant" and "universal" laws of nature. In addition, to doubt science's infallibility was to doubt established theories, in this case mechanical theories. That is, a thoroughgoing skepticism was potentially erosive of atomo-mechanical orthodoxy. Neither Langley nor Nipher fully articulated these two implications of their skepticism, but they did touch on both. As the century drew to a close, physicists would increasingly adopt this skeptical stance and increasingly use it to denigrate traditional mechanical thought.

Langley and Nipher had contrasting educational backgrounds and scientific careers. Samuel Langley (1834–1906), a descendent of prominent colonial families, including the Mathers and Adams, received his formal education at the respected Boston Latin and Boston High Schools.[1] Forgoing college, he began working in the midwest as a civil engineer and architect, a profession he followed until returning to New England at age thirty. After a few years of building telescopes, a lifelong interest, and traveling abroad, he became an assistant in the Harvard Observatory and then assistant professor of mathematics and astronomy at the Naval Academy. In 1867 he moved to Pittsburgh, joining the Western University of Pennsylvania where he served as a professor of astronomy and physics and as Director of the Allegheny Observatory. There for nearly twenty years, he did pioneering work in astrophysics, with particular emphasis on studies of the sun; the bolometer, a device he invented for measuring infrared radiation, aided him in this research. Langley began the final phase of his career in 1887 when he became Secretary of the Smithsonian Institution, a position which was politically prestigious and which demonstrated the esteem given him by his colleagues. Until his death in 1906, Secretary Langley worked for the improvement of American science while continuing his own researches, increasingly pointed toward aerodynamics and flight. Langley was, incidentally, an avid reader; significantly, he read both John Stallo's iconoclastic *Concepts and Theories of Modern Physics* and Charles Peirce's 1877–78 series of articles on the logic of science. Near the turn of the century, he even recommended *Concepts and Theories* to Henry Adams, a distant relative. And about this same time, he wrote to Peirce that he could "still remember with pleasure my reading of your first paper . . . nearly twenty-five years ago."[2]

By comparison, Francis Nipher (1847–1926) neither attended distinguished schools nor achieved international prominence. But he was respected by his American colleagues.[3] Reared in New York, Nipher, at age sixteen, moved west with his family to Iowa City. He acquired his undergraduate degree in 1870 from the State University of Iowa, studying science and mathematics. Continuing at the University as both an instructor in physics and a graduate student, he also earned a master's degree. He became, in 1874, professor of physics at Washington University in St. Louis, a position he held until 1914 when he retired. A driving force behind the St. Louis Academy of Science, Nipher frequently published in the Academy's *Transactions*. His research interests were eclectic: they ranged from electricity, magnetism, optics, and classical mechanics, to terrestrial magnetism, meteorology, and photography. Although he made no major research contributions, he was an active member of the American physics community, serving, in 1891, as AAAS vice-president in charge of Section B–Physics.

One of Samuel Langley's earliest statements of scientific skep-

Francis Nipher. (Courtesy of AIP Niels Bohr Library)

ticism appeared in his 1884 book, *The New Astronomy*. This was a much-reprinted, general summary of the recent transition from "old" positional astronomy to "new" photometric, spectroscopic, and photographic studies of celestial bodies. Langley expressed, by way of epistemological digressions, distrust of "the great principle of the uniformity of the Laws of Nature"; he believed that unexpected departures from these so-called "Laws" were always a possibility. Specifically, he doubted that "Laws of Nature" could "be separated from the laws of the fallible human mind, in which alone Nature is seen." "May we not receive even the teachings of science, as to the 'Laws of Nature,'" he asked, "with the constant memory that all we know, even from science itself, depends on our very limited sensations, our very limited experience, and our still more limited power of conceiving anything for which the experience has not prepared us?" In Langley's opinion, the current generation of scientists agreed with him and, in comparison to their eighteenth-century forefathers, were more "humble" and "diffident of the absoluteness of their own knowledge."[4]

While president of the AAAS during 1887-88, Langley continued to stress the inherent fallibility of science. In his remarks to the Association as incoming president, he encouraged open-mindedness in scientific discussion "in view of our ignorance as to the real nature and causes of things." Moreover, a year later, he devoted his entire address, as retiring president, to rectifying widespread but unfounded stereotypes of science.[5] He began by attacking the common view of scientific progress, "the march of an army towards some definite end." This was the "retrospective view" of the typical textbook writer "who probably knows almost nothing of the real confusion, diversity, and retrograde motion of the individuals comprising the body, and only shows us such parts of it as he, looking backward from his present standpoint, now sees to have been in the right direction." To accurately portray the development of science, Langley preferred the "less dignified" illustration of a "pack of hounds, which, in the long-run, perhaps catches its game, but where, nevertheless, when at fault, each individual goes his own way, by scent not by sight, some running back and some forward; where the louder-voiced bring many to follow them, nearly as often in a wrong path as in a right one; where the entire pack even has been known to move off bodily on a false scent."

As an example of a "louder-voiced" hound leading the pack astray, Langley pointed to the dominance, during the eighteenth century, of Newton's corpuscular theory of light over the undulatory theory. During that century, the corpuscular theory gradually grew "to be an article of faith in a sort of scientific church, where Newton . . . [had] come to be looked on as an infallible head, and his views as dogmas, about which no doubt . . . [was] to be tolerated." This example of scientific orthodoxy led Langley to a "general rule": "that the same thing may appear intrinsically absurd, or intrinsically reasonable, according to the year of grace in which we hear of it." Langley decried the "unsuspected influence of mere tradition in science." And, once again, he emphasized that there is "no infallible guide," "no absolute criterion of truth" for either the individual or community, and that scientific "truths" are put forward "as provisional only." Yet he believed that, over time and "as a whole," science was advancing, not necessarily in any ontological or ultimate sense but in a practical, material sense. His own science of radiant energy, for example, was increasing its domain of application, predictive powers, and technological relevance; from a functional perspective, this science had an "unbounded" future.[6]

A decade before Langley's AAAS speech, Francis Nipher expressed similar skeptical views in a speech to the alumni of the State University of Iowa. In this 1878 speech, later published as a pamphlet, Nipher demonstrated the need for "intellectual modesty" among scientists—and even more so among laymen—by pointing out "the difficulties to which we are subject in arriving at our conceptions of physical law." The first difficulty involved scientists' inevitable "ignorance of certain facts." This ignorance prohibited scientists from establishing absolute laws that predicted events "indefinitely into the future." That is, in any scientific investigation, no matter how seemingly complete, there always remained "many minor disturbances or perturbations, too small to be detected by instruments." A related difficulty in formulating absolute laws was that "events wholly unexpected to our partially instructed minds—apparent breaches of continuity—are liable to happen at any time." Quoting the English philosopher William S. Jevons, Nipher explained that great and sudden "catastrophes" are not beyond the "reign of law"; they are just beyond scientists' under-

[110]

standing of the true law. A further difficulty involved the practical necessity in scientific research of ignoring certain known variables that are negligible but nevertheless relevant. This neglect meant again that physical laws were not complete and exact, but merely the mathematical "first terms" of actually complicated "infinite series." Finally, Nipher believed that physical laws often entailed abstract, nonempirical concepts (such as the ether) that were beyond scientists' "power of realization." And since scientists were "fallible" humans who often made mistakes and errors, the laws built on such "unknown," "strange," and "marvelous" concepts were therefore suspect. Consequently, scientists should be "cautious and modest" in their attitude toward their so-called "final" physical laws, and they should avoid "a slavish acceptance of any theory."[7]

Having summarized the "errors to which scientific men are liable, in arriving at what we provisionally call correct conceptions of physical law," Nipher closed with a few general observations. He surmised that "all scientific results are attended with uncertainty." In those cases, moreover, where the uncertainty is fundamental and encompasses dispute as to what is "truth or error," scientists often become intolerant of the opposing view. Such dogmatism also occurs among laymen in similar cases of uncertainty. In these unsettled situations, Nipher lamented, "we have A making strenuous efforts to convert B and C to his own opinion, failing in which, he proceeds to burn them, to imprison them, to lampoon them in the newspapers, or to do some of the more quiet, but scarcely less effective things, characteristic of our own times, that the spirit of the age will permit." Nipher added, however, that prospects were improving: "Thoughtful men are becoming more and more impressed with the vastness of the unknown, and the comparative insignificance of human achievement, while the demonstrated fallibility of human reason leads them to temperance and modesty of thought and expression; to *appreciation,* as well as toleration, of opposition and doubt. . . . If we feel called upon to defend the truth, we are, after all, only defending what we *believe* to be truth, and possibly against men as honest and able as ourselves."[8]

Having argued that physical science was inherently fallible, what did Langley and Nipher view as the specific implications of this fallibility for established atomo-mechanical theories? For Langley, it implied caution and restraint in accepting, for example, the kinetic theory of gases. In his AAAS address, he suggested a possible parallel between the eighteenth-century history of the phlogiston theory and what might be the future of the modern kinetic theory. The phlogiston doctrine, he explained in words reminiscent of Stallo, was previously "accepted not so much as a conditional hypothesis, [but] as a final guide, and a conquest for truth which should endure always." This doctrine, moreover, "was tested more seriously than the kinetic theory has yet been." Pointing out that the once dominant phlogiston outlook was now completely abandoned, Langley allowed his audience to apply "the obvious moral to hypotheses of our own day." For Langley, the fallibility of science also necessitated reluctance to speculate on the detailed character of heat and light. In both the 1888 address and in the 1884 book, *The New Astronomy*, he stressed that heat and light are not "things in themselves" or "external things" but "merely effects of this mysterious thing we call radiant energy." In assigning the label of "radiant energy" to this "uncomprehended something" or "unknown thing," scientists were doing nothing but giving "a name to the ignorance which still hangs over the ultimate cause."[9]

Nipher also dealt with issues concerning radiant energy when, in 1891, as AAAS vice-president of Section B, he spoke on "The Ether." His skepticism had endured from the time of his Kansas City speech thirteen years earlier. Granting that there seemed to be "important and pressing" needs for an ether, he still could "not help wondering occasionally with Theophrastus Such what kind of hornpipe we are dancing now." He proceeded to draw out and accentuate the adverse implications of the recent writings of Gibbs and Michelson for the ether. Echoing Gibbs's 1888 and 1889 comparisons of the elastic-solid and electrical theories of light, he indicated that the elastic ether theory "was burdened with serious difficulty" as compared to Maxwell's "rival theory." As for Maxwell's own mechanical models involving the ether, Nipher brushed them off as nothing more than heuristic, conceptual aids. In addition, to illustrate how "exceedingly difficult" it was to construct a consistent mechanical model of the luminiferous ether, Nipher pointed to Michelson and Morley's 1887 study of the relative motion of the earth and ether. Recalling their closing suggestion that the earth perhaps had trapped portions of ether, Nipher concluded

[112]

that the relation of ether to matter in motion "is not simplified by the beautiful experiment of Michelson and Morley." Not surprisingly, when Nipher published his own textbook on electricity and magnetism in 1895, it was a mathematical treatise devoid of speculations on the ether and the ultimate nature of electricity.[10]

Langley and Nipher shared with John Stallo an iconoclastic bearing toward aspects of mechanical physics. Stallo himself, nevertheless, held fast to at least one traditional vision: that the overall enterprise of science, as unspoiled by mechanical "dogma," was progressing toward absolute natural laws. Thus, a component of Stallo's ideology was actually closer to Mayer's scientific faith than to Langley's skepticism. During the early 1880s, for example, Stallo assured his readers that he did not in principle doubt "the constancy of physical laws or the universality of their application." He also felt that well-founded hypotheses "are more than mere arbitrary and artificial devices for the enchainment and classification of facts. They are in most cases guesses at the ultimate truth. . . ." During the early 1870s, Stallo's attitude was even further removed from Langley's and Nipher's skepticism. He had expressed a Kantian belief in the "*a priori* sanity of the human intellect" and the "primordial correspondence between the intellect and its object."[11] On the other hand, Langley stressed that so-called laws of nature are actually "laws of the fallible human mind," while Nipher repeatedly called attention to the "demonstrated fallibility of human reason." For Langley and Nipher, physical science in general, and atomo-mechanical physics in particular, were to be approached cautiously and skeptically, with full realization of science's inherent capacity for error.

11

A REAPPRAISAL OF STALLO AND HIS CRITICS

Broadly speaking, atomo-mechanical views prevailed around 1880. We have seen, however, that these views were diversified: there were differences both on the level of substantive content and scientific ideology. American mechanical thought ranged from the orthodoxy of Alfred Mayer and Amos Dolbear, through the cautious reserve of John Trowbridge and Henry Rowland, to the simple experimental practicality of Albert Michelson and Edwin Hall. We have further seen that there existed outlooks that were potentially nonmechanical. Simon Newcomb's operationalism was one; the qualified phenomenalism of J. Willard Gibbs and the scientific skepticism of Samuel Langley and Francis Nipher were others.

Thus, although an atomo-mechanical outlook dominated in the United States, it was neither as uniform nor as firmly implanted as Stallo contended. Stallo oversimplified when he asserted: "With few exceptions, scientific men of the present day hold the proposition, that all physical action is mechanical, to be axiomatic. . . . And they deem the validity of the mechanical explanation of the phenomena of nature to be, not only unquestionable, but absolute, exclusive, and final."[1] In actuality, from about 1870 to 1895, both the content and ideology associated with the ubiquitous mechanical outlook were in a stimulating state of flux and ferment.[2] We can elaborate on this general finding by explicitly answering the four questions that we posed earlier in response to the assertions and counterassertions of Stallo and his critics.

The first question asked whether or not portions of Stallo's critique were passé by the 1880s. In the preceding chapters, we have seen that American physical scientists were in fact aware of technical and metaphysical pitfalls that imperiled the atomo-mechanical theory—pitfalls that Stallo had detailed which, in certain cases, his critics readily had acknowledged. It was an accurate self-appraisal when Stallo mentioned in the preface to *Concepts and Theories of*

Modern Physics that his views were "but the inevitable outcome of the tendencies of modern science . . . the fruits of the epoch rather than of my own invention."[3] Specifically, most scientists were familiar with the technical deficiencies, such as the inability to account for specific heats. They, in contrast to Stallo, did not view these acknowledged failings with urgency or alarm; rather, they treated them as difficult puzzles awaiting routine resolution. Then, too, most scientists were well versed in at least the rhetoric of anti-metaphysical empiricism. At least in principle, they knew the main epistemological and ontological traps to be avoided in scientific research. Whether Americans went beyond this rhetoric and put into practice their antimetaphysical doctrines is an issue we will defer for the moment.

It is not surprising that Stallo's particular list of technical and metaphysical problems did not constitute new and unprecedented knowledge for most physical scientists. After all, Stallo himself had formulated the core of his critique as early as 1873–74 in his *Popular Science Monthly* articles. His own image of physics dated all the way back to the 1840s when he first studied and taught the subject. Stallo's contribution to science was to consolidate, as he said, "a considerable amount of scattered material ready to the hand" into an "orderly and systematic" analysis of modern physics.[4]

The second question dealt with whether American physicists held fundamental atomo-mechanical propositions similar to the four Stallo had listed. We can now answer that they did, but only in a qualified sense. In our survey, Alfred Mayer was the only physicist to *formally* enumerate such propositions. Probably like many other professors, he drew up the list primarily to facilitate instruction. Most of the physicists with mechanical leanings avoided Mayer's type of didactic summary of fundamental propositions. This did not necessarily mean, as Stallo's critics had concluded, that most physicists did not hold such propositions; rather, they often endorsed them in a more offhand, casual, or implicit manner. Moreover, as Stallo also had maintained, they often contradicted these implicit axioms in their actual researches. In other words, according to Stallo's criterion of consistency in theory and practice, they were ambivalent and logically inconsistent in their overall outlooks. Mayer, for example, contradicted his own pedagogic axioms in the course of his regular research activities. Recall that he defended the axiom of noncontact interatomic forces but soon thereafter rejected all actual theories based on action at a distance.

But we must qualify. In truth, except for Gibbs in his formulation of statistical mechanics, the scientists whom we reviewed were not overtly concerned in their daily work with logical perfection or axiomatic purity. The mechanical physics of practitioners like Trowbridge was, by frank admission, functional in design and fluid in form. Consequently, it was especially vulnerable to a philosopher's abstract and static critique. In other words, what appeared from a strictly logical perspective to be inconsistencies, appeared from a workaday research perspective to be the usual dynamics of ongoing scientific inquiry.

We must also carefully qualify our response to the third question. Were American physical scientists, as Stallo had declared, metaphysically committed to the atomo-mechanical theory? Or were these scientists, as Stallo's critics had responded, committed to the theory only provisionally or tentatively? And even if their outlook was avowedly provisional, did it nevertheless entail, as Stallo had further believed, a subtle ontological bias? The answer to this question is neither a simple "yes" or "no." As G. Stanley Hall and other critics of Stallo contended, most American physical scientists disclaimed the ontological significance of detailed atomo-mechanical models and concepts. Among the "mechanical" physicists, Trowbridge, Rowland, Michelson, and Edwin Hall all emphasized the tentative, heuristic, hypothetical, or provisional nature of particular constructs. And, of course, Gibbs, Langley, and Nipher staunchly advocated provisional perspectives. None of these scientists would have asserted, for example, that vortex atoms had ultimate meaning and explanatory power. Even that purported arch-mechanist, Sir William Thomson, counseled ontological restraint with regard to detailed mechanical models in his 1884 message to Americans. And traditionalist Alfred Mayer deviated from his otherwise unquestioning faith by remarking that in the final reckoning science dealt merely with "contingent truths." Perhaps only Amos Dolbear, primarily a science educator and popularizer, openly displayed a naively realistic and hence "metaphysical" attitude toward detailed atomo-mechanical concepts.

But can we take all of these adherents of provisional physics at their word? And can we limit our appraisal to only their attitudes

[116]

regarding the details of atomo-mechanical physics? Mayer, for instance, despite his occasional disclaimers, still insinuated that all scientific knowledge had ontological import. According to Stallo's strict definition, he was "metaphysically" committed to mechanical physics. And whereas Rowland distrusted the details of mechanical physics, he implicitly endorsed a number of its broader concepts, principles, and goals. He was confident of the existence of the ether, though the fine points of this medium remained hazy; and he was sure that a wide range of physical phenomena were explicable through the workings of the ether, though there was little immediate prospect of fully achieving such explication. If we can interpret this diffuse confidence and optimism to imply a philosophical stance on the ultimate nature of physical reality, then Rowland was, in a subtle sense, metaphysically committed to his research program. Although a professedly provisional investigator, he perhaps had unwittingly begun, in Stallo's words, "to treat the fictions . . . as undoubted realities, whose existence no one can question. . . ." [5] Thus, whereas most Americans affirmed that the intricate properties assigned to atoms and ether were merely heuristic aids to research, many were implicitly harboring a subtle metaphysical bias in favor of the broader principles of atomo-mechanical physics. Nonetheless, few Americans would have sanctioned Dolbear's blatant and thoroughgoing mechanical realism.

Finally, there is a twofold answer to the fourth question: were the Americans failing to tie their concepts and theories adequately to the observational realm, or was this merely the misperception of a layman out of touch with actual scientific practice? We have seen, on the one hand, that there was a trend toward operational thought among American physical scientists. On the other, this did not mean that the scientists were in fact operationally evaluating and defining those fundamental atomo-mechanical concepts which Stallo, as a layman, rejected as vague and metaphysical. The writings of Simon Newcomb and also those of Charles Peirce suggest that there was a tendency toward operationalism in the years around 1880. To combat nonempirical, and hence potentially metaphysical thought, these scientists urged critical examinations of language and meaning and they stressed sensate definitions or measurements of individual terms, concepts, or propositions. But neither Newcomb nor Peirce rigorously applied his operational

[117]

critique to the concepts of mechanical physics. Newcomb, in particular, uncritically defended the hypothetical mechanical program, while reserving his operational reprimands for mainly theologians and philosophers. He assumed that modern physics had a healthy empirical constitution and that it had avoided the metaphysical ills still plaguing theology and philosophy.

Of the many American physical scientists who had not consciously enunciated formal operational principles, most assumed, like Newcomb, that atomo-mechanical concepts had empirical warrant. Most were also well versed in the antimetaphysical rhetoric of nineteenth-century empiricism. And most—Michelson in his ether studies and Rowland in his measurements of the mechanical equivalent of heat—were dedicated to routinely expanding the data base of received mechanical theories. But like Newcomb and Peirce, with their formal operational doctrines, these other scientists, with their casual empirical precepts, did not systematically seek to reevaluate and justify the experiential foundations of mechanical physics. Content for the most part with tacit, implicit perceptions concerning empirical roots, American physicists thus were open to the accusation of an outside lay observer, especially a philosopher like Stallo, that concepts such as "ether" were empirically vague and hence potentially metaphysical. Given the shifting pattern over time of what constitutes valid empirical evidence and legitimate scientific theory, these scientists were certainly justified, in a historical sense, in their perception of empirical rigor. Similarly, given the difficulty in actual practice of establishing *direct* correspondences between theoretical concepts and concrete operations, these scientists were also justified, in a philosophical sense, in their perception of empirical rigor. Nevertheless, by the end of the century, scientists increasingly would be both reinventing Stallo's largely forgotten critique and consciously implementing latent operational doctrines. These events would contribute to the faltering of the hitherto dominant, but notably diversified, atomo-mechanical outlook.

CONCEPTUAL FERMENT AND
REVOLUTIONARY EXPECTATIONS

Carl Barus of Brown University, ca. 1900. (Courtesy of AIP
Niels Bohr Library)

12

REALIGNMENTS WITHIN THE OLD GUARD

In 1904, exactly twenty years after speaking at the Johns Hopkins University, William Thomson belatedly published his *Baltimore Lectures on Molecular Dynamics and the Wave Theory of Light*. Whereas in 1884 American physicists generally thought well of the course of lectures, they received the 1904 volume with reserve, perhaps even embarrassment. Joseph Ames, writing in the *Physical Review*—the ten-year-old, professional journal of American physicists—endeavored to criticize the book tactfully without tarnishing the sterling image of its venerable author. A professor at Johns Hopkins and a protégé of the now deceased Rowland, Ames spoke forbearingly of "difficulties" in reviewing this "important" volume. A major problem was that "within the past ten years physics has developed along lines essentially different from those which are treated in this volume, and therefore one lacks that interest which would have been aroused if it had appeared twenty or even fifteen years ago." "One notes with regret," Ames added, "the absence of reference to many recent investigations which have important bearing on the subjects discussed."[1]

The changes that led to the obsolescence of Thomson's *Baltimore Lectures* were dramatized later in 1904 when many of the world's principal physicists assembled in St. Louis. The occasion was the Congress of Arts and Science, held in conjunction with the St. Louis Universal Exposition. The United States had never hosted a more important gathering of physicists, not even during the 1884 meetings of the Philadelphia Electrical Conference and the Baltimore Lectures. Thomson had dominated the 1884 assemblies, outshining British and American leaders such as Rayleigh and Rowland, but now he remained home in England—an eighty-year-old scientific laureate with the honorary title of Lord Kelvin.[2] In his stead, there journeyed to St. Louis a younger generation of international standard-bearers whose names today connote scientific upheaval: Henri Poincaré, Paul Langevin, Ernest Rutherford, and

Wilhelm Ostwald. These men shared the St. Louis spotlight with a similarly progressive but less well-known generation of Americans including Edward Nichols, Carl Barus, Arthur Kimball, Dewitt Brace, Robert Millikan, and Henry Crew. By 1904, both Thomson and his *Baltimore Lectures* were merely esteemed monuments to an earlier age.

Ames specified in 1904 that Thomson's molecular dynamics had lost relevance "within the past ten years." Like many of his American and European colleagues, he dated the turmoil which was occurring in world physics from about the latter part of 1895 when Wilhelm Roentgen discovered X rays. Roentgen's X rays, accidentally detected, and having unanticipated penetrative properties, caused a sensation among scientists. Henri Becquerel's subsequent search for X rays in fluorescence phenomena also carried him, in 1896, to the equally fortuitous and unsettling discovery of radioactivity. This in turn quickly led to evidence for atomic transmutation with the work of Marie and Pierre Curie, Ernest Rutherford, and Frederick Soddy. Meanwhile, J. J. Thomson's experimental characterization of the electron in 1897 compounded the general excitement of this period.

But American physicists, like their European counterparts, perceived this era to be more than one of novel experimental results. They were aware of associated unrest which was also occurring in the realm of theory. Specifically, there was a vibrant interplay of the traditional atomo-mechanical, statistical-mechanical, thermodynamic, and electromagnetic views of physical phenomena—the latter three views now generally recognized as viable research alternatives. Although a number of Americans still espoused the traditional mechanical outlook, it was increasingly being challenged and criticized, often on explicitly operational or positivistic grounds. Moreover, certain mathematically adept physicists hoped to subordinate orthodox atomo-mechanics to generalized statistical mechanics; and scientists who advocated the more phenomenological outlook of Wilhelm Ostwald hoped to install thermodynamics as the cornerstone of physical explanation. Other Americans sought an alternative to orthodox mechanism by espousing electromagnetic theory, particularly as presented by H. A. Lorentz. (In reviewing Thomson's *Baltimore Lectures,* Ames marveled that "there is no mention made of the theories of Lorentz.") Scientists

[122]

holding the extreme of this electromagnetic view, as championed by Europeans Wilhelm Wien and Max Abraham, hoped to reduce all of classical mechanics to electromagnetic concepts.

Turmoil also existed in the realm of scientific ideology. American physicists were increasingly abandoning the already tattered remnants of their belief in an omniscient science that was making steady progress in revealing absolute natural laws. Instead, physicists were resigning themselves to metaphysically neutral descriptions or correlations of phenomena. Atomo-mechanical confidence was giving way to sober skepticism and agnosticism toward both the means and ends of scientific inquiry.[3]

Although Roentgen's 1895 discovery of X rays signalled the onslaught of an intense period of experimental, theoretical, and ideological ferment, we must remember that the ferment derived from events of the preceding two decades. Even while William Thomson was lecturing in Baltimore in 1884, the dominant atomo-mechanical outlook was branching into an array of varied perspectives that would be fruitful for internal competition and change. Additionally, scientists were recognizing definite failings of the mechanical research program. Furthermore, flourishing alongside mechanical physics were three fresh options. The emerging thermodynamic, statistical-mechanical, and electromagnetic outlooks promised the type of logical consistency and explanatory power that Stallo had found lacking in traditional mechanical physics. Finally, through its growing identification with a provisional, empirical, and antimetaphysical ideology, the mechanical viewpoint was nurturing the seeds of its own demise. That is, the "tentative" details of mechanical hypotheses would be vulnerable to the skeptical and operational forces already solidifying in the decades bracketing 1880. The discovery of X rays would merely signal the quickening of all of these prior movements.

To illustrate this heightened conceptual ferment and surge of revolutionary expectation, let us return to those physical scientists who were active from 1870 to 1895 and look for shifts in their outlooks during the subsequent decade. It turns out, for example, that Rowland and Trowbridge wavered in their mechanical views; Langley hardened his skepticism; and Gibbs became even more shy of detailed mechanical assumptions. On the other hand, some of the physicists earlier surveyed sought, through retrenchment, to

[123]

protect threatened mechanical positions. Francis Nipher was one of these.

The changes in Henry Rowland's outlook evident after 1895 were probably typical of those made by American physicists with previously moderate mechanical leanings. These changes were subtle, best described as new emphases rather than dramatic transformations. Recall, for example, that Rowland, although aware of problems with the ether, was committed, nevertheless, to its central, unifying role in physics. Quite correctly, upon Rowland's death in 1901, a colleague concluded that the "fixity of his conviction" regarding Faraday's and Maxwell's view of molecular mechanics and the ether had excluded him from taking part in recent "modifications and development" of electromagnetic theory such as made by Lorentz.[4] But while he continued to endorse this traditional ether concept through the late 1890s, his commitment was shaken with Roentgen's discovery of X rays. In informal remarks made in 1896 on "The Röntgen Ray, and Its Relation to Physics," Rowland was generally pessimistic that the concept of an all-embracing ether could be reconciled with the phenomenon of X rays. Using words reminiscent of Stallo fourteen years earlier, he began by lamenting the ether's inability to account simultaneously for even traditional, pre-X ray phenomena:

> Now, you have got all those things—electromagnetic action, light which would be an electromagnetic phenomenon, and then we have gravitation, and we have got to load the ether with all those things. Then we have got to put matter in the ether and have got to get some connection between the matter and the ether. By that time one's mind is in a whirl, and we give it up.
> Now we have got something worse yet—we have got Röntgen rays on top of all that. Here is something that goes through the ether, and it not only goes through the ether but shoots in a straight line right through a body. Now, what sort of earthly thing can that be?[5]

Rowland concluded that physicists had little immediate hope of understanding X rays and their relation to the ether.

Rowland's heightened doubts about the ether correlated with subtle shifts in his scientific ideology. These shifts, in turn, perhaps reflected the onset during the 1890s of a tragic personal problem:

[124]

his awareness that he was dying of incurable diabetes. Although he had always been resigned to temporary gaps in scientific knowledge and had avoided speculations about the detailed workings of phenomena, he previously had remained confident of science's ultimate ability to discover absolute natural laws. However, in an important speech in 1899, less than two years before his death, he advanced an ideology closer to Langley's earlier skepticism. Because he was speaking both as the first president of the newly founded American Physical Society and as one of the most eminent American physicists, his comments were widely reprinted in such journals as *Science,* the *American Journal of Science,* and the 1899 inaugural issue of the *Bulletin of the American Physical Society.*[6] Cautioning his audience of the limitations of "imagination unguided by experiment" (regarding recent speculations on charged particles moving with speeds comparable to that of light), Rowland offered his "law of the conservation of knowledge": in any "mathematical investigation . . . we never get out more from it than we put in." (During the next few years, American physicists would frequently quote this aphorism.[7]) Moreover, he continued, given "the liability to error in whatever direction we go, the infirmity of our minds in their reasoning power, [and] the fallibility of witnesses and experimenters," a scientist should always "be specially skeptical with reference to any statement made to him or any so-called knowledge which may be brought to his attention."

On this latter point, Rowland became more emphatic: "There is no such thing as absolute truth and absolute falsehood. The scientific mind should never recognize the perfect truth or the perfect falsehood of any supposed theory or observation. It should carefully weigh the chances of truth and error and grade each in its proper position along the line joining absolute truth and absolute error." He went on to conclude: "The ideal scientific mind, therefore, must always be held in a state of balance which the slightest new evidence may change in one direction or another. It is in a constant state of skepticism, knowing full well that nothing is certain. It is above all an agnostic with respect to all facts and theories of science as well as to all other so-called beliefs and theories." Skepticism and agnosticism had tempered his earlier faith in the "true and overwhelming progress of science which marches forward to the understanding of the universe."

John Trowbridge had also recently begun to use the word agnosticism. Seeking to avoid the negative connotations of the word, Trowbridge explained in 1896 that "agnosticism in physical science is a hopeful creed when it is enlivened by a quick imagination." He elaborated:

> There is no tendency to restrain the imagination in this attitude of scientific agnosticism. The physicist of to-day has his ethers and his atoms just as the ancient Greek and Roman philosophers had theirs, and he pictures to himself invisible motions far more subtle than entered the imagination of Aristotle or Democritus. The natural philosopher of to-day, however, differs in this essential respect from the ancient philosopher: he measures. If his heat measures [based on the principle of conservation of energy] do not agree with his hypotheses of vortical or atomic motions, he rejects his attractive hypotheses instead of hugging them.

Trowbridge made this declaration in the introduction to his new volume in Appleton's "International Scientific Series"[8]; the volume bore the identical title and covered the same topics as his shorter, 1884 article, "What Is Electricity?"

A comparison of these sister publications reveals that Trowbridge's 1896 stress on agnosticism as buttressed by imagination and measurement was a more precise and stringent restatement of a similar 1884 emphasis on the unknowableness of ultimate scientific answers. A comparison also shows that Trowbridge in 1896 spelled out the problems confronting ethereal and atomistic hypotheses, an analysis not evident in the earlier article. Putting into practice his formal agnostic precepts, he forthrightly explained that the vortex theory of the atom was a "forced theory." Furthermore, in 1884, he had optimistically prophesied a grand atomomechanical unification of electricity, magnetism, light, heat, gravitation, and chemical force; by 1896 he had resigned himself to a more modest and limited synthesis. In particular, while still occasionally speculating on the atomic and ethereal motions underlying gravitational and chemical processes, he focused mainly on the better established and more tractable mechanical connections between light, heat, and electromagnetic waves.[9]

Incidentally, in one of his passing conjectures on chemical phenomena, Trowbridge made what was for American physicists a rare reference to Stallo's *Concepts and Theories of Modern Physics,* also a volume in Appleton's "Series." It was, however, an ironic reference. To demonstrate the mechanical nature of electrochemical forces, Trowbridge cited a German investigation originally included by Stallo to show the futility and absurdity of mechanical explanations of chemical action.[10] Such continued atomo-mechanical conviction on Trowbridge's part, while mitigated since 1884, probably contributed to the negative reception of his book. A critic for the *Physical Review* found *What Is Electricity?* to be plagued not only by "omissions" of relevant material but also by numerous "half-truths, misleading statements, and errors." Nevertheless, by 1904, with the publication of later writings, sixty-one-year-old Trowbridge would demonstrate the sincerity of his scientific agnosticism and open-mindedness by moving further from traditional nineteenth-century modes of thought toward the newer electron theory.[11]

While Rowland and Trowbridge liberalized their prior scientific ideologies by openly embracing agnosticism, Samuel Langley, already a longstanding skeptic, grew more adamant in his doubt. He revealed his hardened outlook in a speech given in 1902 before the Washington Philosophical Society. Titled "The Laws of Nature," the speech appeared later as the lead article in *Science* and as an entry in Langley's own *Smithsonian Reports.* In his earlier years, he had been tentative and guarded in his misgivings about natural law, lacing his occasional philosophical digressions with qualifications. By 1902, his assertions were stark and unequivocal. Joining current European philosopher-scientists Ernst Mach and Karl Pearson, he stated: "It is perhaps a hard saying to most that there are no such things as 'laws of nature'; but this is the theme on which I have to speak." Natural laws were simply "the laws of man's own mind, or the effects of his own mind, which he projects outside of himself and imagines to be due to some permanent and unalterable cause having an independent existence." Langley granted that scientists may know a little of the "order" of nature, but he was emphatic that they could know nothing of the "laws." To speak of "an observed 'order' of nature" did not carry with it the groundless "implication of necessity denoted by 'law.'"[12]

To this stark skepticism, Langley added another new element—cynicism. Not only did he doubt the possibility of obtaining ultimate knowledge of nature, he now openly distrusted and disparaged the motives of traditional scientists. In a misanthropic and anti-deistic tone, he listed reasons why a man would presume to fathom absolute laws of nature:

> This is not only because his season for observation is but a moment in the passage of nature's eternal year, and because of his pathetic sense of his own weakness he would gladly stay himself on the word of some unchanging being. It is because this sense of dependence is strangely joined with such self-conceit that when he listens to what he himself says he calls it the voice of God. From these twin causes, arising both from his inability as a creature of time to observe nature, which is eternal, and again from his own overweening sense of his own capacity to know her, he looks for some immutable being whom he believes to have written his own ideas in what he calls "the book of nature."

Building on these themes of human weakness and conceit, Langley reiterated his earlier conclusions regarding the transitory and provisional character of scientific theories, including even the kinetic theory of gases.[13]

Langley did grant in this speech, as he had hinted in earlier years, that attitudes toward natural law were changing. Even though the phrase "laws of nature" continued to exercise a "wide influence," he sensed that "a significant change is taking place in the leaders of common opinion with regard to the meaning that the words convey." In studying nature, scientists "are no longer impressed by her 'laws' as were the scientific men of a former generation." He added that this "great change" had begun to take place within the period of his own personal recollection. In a letter to Charles Peirce written in 1901, Langley summarized the temper of the transition: "To the common of educated men a 'law of Nature,' I think, is now coming to mean a statement of an *apparent* sequence in Nature, which is only important as far as it has been correctly observed and interpreted by the fallible human mind."[14] While skeptical of absolute natural laws and cynical toward those who believed in them, he hoped that science was entering, as he stated in his 1902 speech, a more "modest" and "humble" age.

Langley's comments to Charles Peirce on natural law were not meant as polite discourse in a sociable letter. Rather, Langley,

Secretary of the Smithsonian, was generously attempting to encourage and guide the unemployed and indigent Peirce to write a saleable article on changes in views toward natural law. Unfortunately, Langley felt that Peirce's subsequent manuscript was too abstruse both in language and content. The correspondence ended when Langley himself wrote "Laws of Nature" in 1902.[15] Nevertheless, one wonders what Peirce's mature views on natural law and physical science were.

Peirce, while younger, had implicitly accepted many of the conventional precepts of atomo-mechanic physics. By 1887, however, he had begun to speak of "the breakdown of the mechanical philosophy" and the "doom" of the associated "necessitarian metaphysics." And by 1891 this philosopher-scientist was firm in his antimechanical stance. Aware of technical failings such as the inability of the kinetic theory of gases to account for specific heats, he concluded that "there is room for serious doubt whether the fundamental laws of mechanics hold good for single atoms." On a more general level, the possibility of statistical violations of the second law of thermodynamics also led him to mistrust the mechanical philosophers' belief in "the precise and universal conformity of acts to law." He consequently turned from a deterministic, necessitarian view of natural law to an "evolutionary" view—a view that some present-day commentators feel is consonant with one that later accompanied quantum mechanics:

> Now the only possible way of accounting for the laws of nature and for uniformity in general is to suppose them results of evolution. This supposes them not to be absolute, not to be obeyed precisely. It makes an element of indeterminancy, spontaneity, or absolute chance in nature. Just as, when we attempt to verify any physical law, we find our observations cannot be precisely satisfied by it, and rightly attribute the discrepancy to errors of observation, so we must suppose far more minute discrepancies to exist owing to the imperfect cogency of the law itself, to a certain swerving of the facts from any definite formula.[16]

Langley, a thoroughgoing skeptic, had found the concept of natural law meaningless; Peirce, still a philosophical realist with trust in the potential of the human mind, retained the concept of natural law but fundamentally redefined it.

[129]

J. Willard Gibbs, mid-1890s. (Courtesy of AIP Niels Bohr Library)

[130]

Near the turn of the century, Peirce also self-consciously placed pragmatism near the center of his philosophy: he began to refine and emphasize his hitherto undeveloped philosophical perspective. This followed William James's 1898 rediscovery and popularization of Peirce's original *Popular Science Monthly* articles.[17] For the most part Peirce still viewed himself as bringing the operational methods of science to the aid of philosophy. On occasion, however, he did explicitly direct his updated critique at scientists themselves. He no longer implicitly defended the atomo-mechanical program as he and Simon Newcomb had done twenty years earlier. For example, he brought the full brunt of his operationalism to bear on an attempted reformulation of mechanical physics made in 1898 by Silas Holman, emeritus professor of physics at the Massachusetts Institute of Technology. In an 1899 review of Holman's book, *Matter, Energy, Force and Work,*[18] Peirce maintained that Holman had neglected a basic "maxim of logic": "the meaning of a word lies in the use that is to be made of it, so that every term of general physics ought to stand for a definite general phenomenon; and whoever clearly apprehends to what phenomenon a physical term refers, has nothing further to learn about that term except its grammatical construction." He particularly disapproved of Holman's redefinition of basic terms like weight and kinetic energy. In Peirce's opinion, "to introduce distinctions of terminology which refer to no differences in the phenomena, is an idle pedantry. . . ." Among other misgivings, he also questioned Holman's devotion to the vortex theory of the atom. Holman had continued to endorse the ether-based theory of vortex atoms while at the same time citing a recent letter from Lord Kelvin that argued against the theory. Kelvin expressed "regret" that he no longer believed it possible "to explain all the properties of matter by the Vortex-atom Theory alone."[19] For Peirce, Holman's unremitting defense of vortex atoms that were joined to each other through an ethereal medium, was merely a misguided attempt to avoid the phenomenon of action at a distance by means of "*a priori* metaphysics."

Around 1900, J. Willard Gibbs put into practice one of the central attitudes shared by colleagues Rowland, Trowbridge, Langley, and Peirce: a diminished confidence in attempts to detail absolute mechanical laws. This attitude, which colored Gibbs's studies throughout his career, emerged as a fullblown doctrine in his in-

[131]

fluential textbook, *Elementary Principles in Statistical Mechanics: Developed with Especial Reference to the Rational Foundation of Thermodynamics* (1902).[20] In this book, which was the culmination of two decades of prior thought, Gibbs approached statistical mechanics as an exercise in abstract mathematics and general Hamiltonian mechanics, independent of specific physical or atomic assumptions. Even though statistical mechanics was historically rooted in thermodynamic phenomena, it was "eminently worthy of an independent development." Only in the final three chapters of this fifteen-chapter text did he draw "analogies" between his abstruse Hamiltonian equations and concrete thermodynamic processes. Moreover, he employed new statistical methods that were also exceptionally abstract and general. Instead of accepting the traditional statistical analysis of a large number of *particles* (e.g., molecules) treated as individuals in one isolated system (e.g., a gas), he preferred "to take a broader view of the subject." He bypassed detailed speculations on the component particles of the system by opting for a statistical analysis of a large number of independent *systems* (e.g., replicas of a gas as a whole considered in different internal states).[21]

Gibbs's abstention from detailed atomo-mechanical assumptions was part of a conscious strategy. In the preface to *Elementary Principles in Statistical Mechanics,* he explained that

> we avoid the gravest difficulties when, giving up the attempt to frame hypotheses concerning the constitution of material bodies, we pursue statistical inquiries as a branch of rational mechanics. In the present state of science, it seems hardly possible to frame a dynamic theory of molecular action which shall embrace the phenomena of thermo-dynamics, of radiation, and of the electrical manifestations which accompany the union of atoms. Yet any theory is obviously inadequate which does not take account of all these phenomena. . . . Certainly, one is building on an insecure foundation, who rests his work on hypotheses concerning the constitution of matter.

On a more personal level, he added:

> Difficulties of this kind have deterred the author from attempting to explain the mysteries of nature, and have forced him to be contented with the more modest aim of deducing some of

the more obvious propositions relating to the statistical branch of mechanics. Here, there can be no mistake in regard to the agreement of the hypotheses with the facts of nature, for nothing is assumed in that respect. The only error into which one can fall, is the want of agreement between the premises and the conclusions, and this, with care, one may hope, in the main, to avoid.

Gibbs's modesty regarding explanations of "the mysteries of nature" persisted into the final three chapters of his book where he turned to thermodynamic "analogies," or the "rational" and "*a priori*" foundation of thermodynamics. Even though his earlier chapters of abstract mathematics now paid off handsomely in derivations of the basic laws of thermodynamics, he again advocated caution: "we are still far from having explained the phenomena of nature with respect to these laws. For, as compared with the case of nature, the [statistical-mechanical] systems which we have considered are of an ideal simplicity." This oversimplification was manifested, Gibbs suggested, in the inability of his statistical methods to account for certain phenomena recognized by physicists to be intractable. The abstract mathematical formalisms, for example, could explain neither the observed values of specific heats of gases nor the phenomena of radiation (particularly radiant heat). Although Gibbs, who died in 1903, did not live to realize it, he had run up against two puzzles that could be solved only by using Max Planck's 1900 hypothesis of the quantum. Thus, his cautious, general, and otherwise successful statistical recasting of the atomo-mechanical program represented one of the limits to which this program in its traditional form could be carried. When later modified to take account of Planck's quantum hypothesis, however, Gibb's statistical methods would prove general enough to flourish in conjunction with modern quantum theory.

[133]

13

RETRENCHMENT

It would be inaccurate to suggest that all the physical scientists active from 1870 to 1895 liberalized their outlooks during later years. For every traditionalist like Rowland or Trowbridge who modified his view around 1900, there was a Michelson, Dolbear, Hall, or Mayer who did not. And for every previously innovative thinker such as Peirce or Gibbs who refined and developed his outlook, there was a Nipher or Newcomb who either relinquished or let atrophy his earlier distinctive perspective. A more balanced image of the conceptual climate takes shape if we follow the thoughts of each of these more conventional thinkers into the later period. Let us begin with Michelson, the scientist who would become, in 1907, America's first Nobel laureate.

Michelson was indeed a physicist who persisted in his initial, conservative patterns of thought; he continued to tie his research to orthodox atomo-mechanical precepts. In 1897, for example, he repeated his earlier interferometer experiments to detect the relative motion of the earth and ether. Suspecting that the prior experiments had failed because the ether was trapped by the earth's irregular surface, Michelson constructed a fifty-foot-high vertical interferometer to search for "a relative motion corresponding to a difference of level." When the modified apparatus gave negative results, he still did not doubt the existence of the ether but merely the existing theories. Michelson's abiding confidence in orthodox mechanical precepts also colored his initial reaction to X rays. Unlike Rowland, who despaired of reconciling X rays with an already overburdened ether, Michelson, in 1896, enthusiastically proposed an "ether-vortex" theory of X rays. In one of his rare formulations of a theory—or, more accurately, a "working hypothesis"—he explained the properties of X rays by supposing them to be "vortices of an inter-molecular medium."[1]

Michelson's traditional attitudes also endured in his more popular writings. This was especially true in a series of eight lectures on "Light Waves and Their Uses" that he delivered in 1899 at the

Lowell Institute and later published in 1903 as a book. Again, in contrast to the scientific agnosticism and skepticism of Rowland and Langley, Michelson began the lectures with an optimistic affirmation of his orthodox faith in the omniscience of science and the orderliness of nature. "What can surpass in beauty," he asked, "the wonderful adaptation of Nature's means to her ends, and the never-failing rule of law and order which governs even the most irregular and complicated of her manifestations? These laws it is the object of the scientific investigator to discover and apply." In his second lecture, he explained that new discoveries of such laws would hereafter come mainly through "extreme refinement in the science of measurement." Confident that "the more important laws and facts of physical science have all been discovered," he believed that "future discoveries must be looked for in the sixth place of decimals." This assertion was as much a rationalization of his own precision studies involving light waves—the topic of his eight lectures—as it was, in the words of a recent historian, a statement of belief in "the completeness of nineteenth-century science."[2]

In these Lowell lectures, Michelson also continued to display an allegiance to the substantive content of mechanical physics. Although conversant with recent innovations like Lorentz's electron theory, he remained most comfortable with strictly mechanical modes of thought, particularly as espoused by Thomson in his 1884 Baltimore Lectures. Echoing the ideas of this old acquaintance and advisor, Michelson favored an "elastic-solid" ether that had its analogue in "shoemaker's wax." And like Thomson (but at variance with scientists like Rowland and Gibbs), he held that the electromagnetic theory of light "is in no sense an explanation of the nature of light. It is only a transference of the problem, for the question then arises as to the nature of the medium and of the mechanical actions involved in such a medium which sustains and transmits these electromagnetic disturbances." He also clung to Thomson's former, "promising" hypothesis of the "ether-vortex." Indeed, in general, Michelson viewed the ether as the key to a possible unification of all physical science:

Suppose that an ether strain corresponds to an electric charge, an ether displacement to the electric current, these ether vortices to the atoms—if we continue these suppositions, we arrive at

[135]

what may be one of the grandest generalizations of modern science—of which we are tempted to say that it ought to be true even if it is not—namely, that all the phenomena of the physical universe are only different manifestations of the various modes of motion of one all-pervading substance—the ether.

For traditionalist Michelson, this synthesis was the goal toward which "all modern investigation tends."[3]

Michelson was not alone in his sustained mechanical aspirations. Amos Dolbear published two books in 1897 extending his earlier doctrines. Although his tone was now less dogmatic and self-assured, he still believed in a physics built solely from the concepts of matter, ether, and motion. A reader could infer Dolbear's general thesis simply by examining the cover of the first of his 1897 books. Above the title, *Modes of Motion,* appeared a labeled illustration of a "coal pile" next to a "boiler" connected to a "steam engine" connected to a "dynamo" connected to an "arc lamp" that gives off "light." If the message of this line drawing in conjunction with the title was not self-evident, Dolbear spelled it out in the text of the book. Beginning with the burning coal, which involved the "molecular motion called heat," all the components of the illustration up to the arc lamp were linked by "motions of ordinary matter"; at the arc lamp "the motions are handed over to the ether, and they are radiated away as light waves." Earlier in the text, he had suggested that the atoms and molecules of "ordinary matter" are themselves "minute vortex rings of ether" or, in other words, "a form of motion of the ether." Dolbear expounded a similar view in his *First Principles of Natural Philosophy,* a book also published in 1897. When later reviewing this textbook, a critic for the *Physical Review* spoke of "the very conservative standpoint of the author." Noticing Dolbear's dependence on vortex-atom hypotheses, the critic added that such speculations "justify fully" Dolbear's return "to the old name, 'Natural Philosophy,'" in the title of the book.[4] Indeed, at sixty, Dolbear was one of America's last natural—or, to be more precise, mechanical—philosophers.

Edwin Hall also carried his tested patterns of thought into the twentieth century. Ill-at-ease with the subtleties and entanglements of theory, mathematics, and philosophy, he remained primarily a skilled experimenter working within a reliable framework of simple mechanisms and concrete analogies. Thus, in a 1904 address as AAAS vice-president in charge of physics, he outlined various analogies between a thermo-electric current and the actions of water in an ordinay steam engine.[5] Just before his death in 1938 at eighty-three, Hall reminisced about his earlier professional career. His recollection of an elderly Harvard colleague, Joseph Lovering, who had objected to updating an outmoded textbook, led him to comment: "I am somewhat shocked to see how much like his attitude my own now is regarding some of the modern developments of physics." Irrespective of these important "modern developments," he believed that he had contributed to an earlier major transition in physics by clearly discerning, for example, the difference between energy and force. Remembering that one of his first students at Harvard in the 1880s later thanked him for teaching the meaning of energy, Hall remarked: "Now that was something worth while, for at that time intelligent people were still talking about the 'conservation of force,' with all the mental confusion that phrase would indicate."[6] In Hall's opinion, it was a notable achievement to clarify and order the basic concepts of nineteenth-century physics.

Not all veteran mechanical scientists maintained or realigned their outlooks during later years. There was a third response to the new intellectual trends. Alfred Mayer simply stopped doing research. After his most productive decade of study, Mayer, around 1880 when he was forty-four, sold his personal library, purchased a country home within commuting distance of Stevens Institute, and terminated his active research career. His son later speculated that Mayer had become lost in the shuffle of the newer physics. Acoustics, Mayer's main experimental field, "had well nigh become a finished subject"; consequently, in 1897, Mayer "died saddened by the thought that . . . [he] had outlived contemporaneous interest in . . . [his] field of study." Actually, Mayer was not so tragic a victim of scientific change. Once ensconced in his country home, he not only continued teaching but also became one of America's leading sportsmen, respected for his shooting, fly casting, and canoeing. His scholarly penchant also persisted. During the 1880s he published such articles as "On the Invention of the Fishing Reel" and "Bob White, the Game Bird of America."[7] Mayer may have died a frustrated physicist, but no one could say he died an unfulfilled fisherman.

[137]

Francis Nipher *(right)* at the St. Louis Congress. From *Popular Science Monthly,* 1904. (Permission of Times Mirror Magazines, Inc.)

In their younger years, Michelson, Dolbear, Hall, and Mayer had all held traditional mechanical outlooks. The former three carried these outlooks into the later years of their careers. Francis Nipher and Simon Newcomb, however, originally had held views that were potentially corrosive of mechanical physics. Nevertheless, in their later years Nipher and Newcomb closed ranks with mechanical thinkers like Michelson, Dolbear, and Hall. Specifically, Nipher backed off from his prior skepticism, and Newcomb never did apply his operational critique to orthodox mechanical concepts.

Nipher's reaction to the turn-of-the-century proliferation of experimental and theoretical novelties was retrenchment, an attempt to protect threatened mechanical positions. He abandoned his prior scientific skepticism and sought to subsume new phenomena and concepts such as radioactivity and the electron under existing mechanical patterns. This defensive posture colored his 1904 speech on "Present Problems in the Physics of Matter," delivered to the St. Louis Congress of Arts and Science. He had in 1878 pointed to strange and unexpected events in nature to illustrate the difficulties in formulating absolute natural laws, but by 1904, he focused on these same types of troublesome events to assure his audience of their amenability to established laws. Building on an analogy popularized by Charles Babbage, Nipher specifically imagined the construction of a calculating machine that for many years would print out an orderly arithmetical series but at some prearranged date would produce "one single arbitrary term." An unsuspecting scientific investigator who examined the machine and worked out the mathematical law of the orderly series would be surprised by the machine's one instance of "seemingly lawless behavior." Nevertheless, Nipher trusted that the scientist would feel "that the principles of mechanism have not been shaken by this unlooked-for disclosure." Rather, taking a broadened perspective, the scientist would begin again his study of the machine "along precisely the same lines, and by the same methods as his previous work." Trusting prior procedures, he would seek the mechanism that had produced the singular term.[8]

The response of this imaginary scientist to the errant calculating machine was similar to Nipher's response to both radioactivity and the electromagnetic view of nature. He did not believe that "old ideas must be renounced in order to explain some new phenomenon which is apparently out of harmony with the explanation previously made." For Nipher, radioactivity was not an unprecedented phenomenon. It was merely a type of "explosion" that happened to display normal explosive properties "to a very exalted degree." Convinced of the identity of radioactive bodies and explosives, he even described some of his as yet unsuccessful experiments using gunpowder "to obtain X-rays from explosions." In like manner, Nipher objected to replacing the seasoned mechanical view of nature with the electromagnetic view. "There seems to be a marked tendency towards the idea that matter and its properties are alike electrical phenomena," he conceded. But he emphatically denied the independent and primitive character of electric charge: "There is nothing whatever to show that electricity has ever been separated from something which has what we have been accustomed to call mass." Nipher was most at ease with the physics of the nineteenth century.[9]

In his later years, Simon Newcomb continued to couple an operational outlook with an unquestioning belief in atomo-mechanical physics, never rigorously applying his potentially corrosive operational critique to traditional mechanical concepts. His operationalism reemerged in 1904 when he, at sixty-nine, served as president of the St. Louis Congress. In a *Popular Science Monthly* article announcing the upcoming Congress, Newcomb anticipated that the gathering would help to unify the various branches of science. "As we go deeper into all the laws of nature," he wrote, "we are led nearer and nearer to the belief that the fundamental principles on which her operations are carried on may be few in number." He did, however, acknowledge an obstacle to this unification. "It may be that our hope of doing anything of the kind has received a great set-back by the iconoclastic way in which the discovery of radio-activity has shaken to its foundations what, ten years ago, were supposed to be fundamental principles at play in the natural world." But Newcomb was generally hopeful that diverse fields would become increasingly integrated, particularly if the practitioners in the fields adhered to sound operational doctrines. In words recalling his earlier writings, he spoke of

> the natural tendency of every science, when pursued by the best methods, to become more precise in the expression of its laws,

and thus to bring mathematical conceptions to the aid of its investigators. When we have not only assigned a name to an object of study, but have made measurement of its size, or of the intensity of any ascertained properties it exhibits, we have taken a great step toward giving precision to our results, and making them comprehensible to a wider body of investigators.[10]

While continuing to make operational statements, Newcomb was still implicitly endorsing the basic mechanical tenets of nineteenth-century physics. For instance, he revealed in one of his most imaginative projects, a science-fiction novel, that he maintained a belief in an elastic-solid ether. In this futuristic novel written in 1900, Newcomb traced the adventures of a "Professor of Molecular Physics in Harvard University" who discovers, in the far-off year of about 1941, a remarkable flight-inducing substance called "etherine." What is etherine? It is, the professor explains, "a new form of matter having relations to the luminiferous ether, not possessed by any other matter formerly known to men." According to "an eminent British mathematician and philosopher, W. K. Constant"—Newcomb's fictional version of Lord Kelvin—etherine "exerts a new reaction upon the ether when made to vibrate in a certain way"; consequently, a small chunk of etherine can "fly through space, carrying any burden whatever as lightly and easily as a bird flies through the air." Ultimately, the Harvard professor uses the etherine to produce a fleet of spaceships with which he subdues the armies of the world and hence insures international peace. Just as Newcomb evolved his scientific fantasies from the basic principles of nineteenth-century ether physics, he also designed his spaceships in accordance with traditional technology. The spaceships were made of oak and the etherine propelling them was activated by burning massive quantities of coal.[11]

14

THE ST. LOUIS CONGRESS: FOREIGN DELEGATES

"On account of its comprehensiveness of plan, the large attendance of foreign scholars of the first eminence, and the picturesqueness (in several senses) of its attendant circumstances, the Congress of Arts and Science of the St. Louis Exposition was doubtless the most memorable and impressive scientific gathering ever held in America." So began Arthur Lovejoy's review, in *Science*, of the published proceedings of the Congress, which convened in 1904. Reviewing the same volume for *The Nation*, Charles Peirce started off by saying, "In every history of the human mind the single success of the St. Louis Congress must be commemorated." The immediate reaction of those attending the Congress had also been enthusiastic. In the lead article in *Popular Science Monthly* a few weeks after the Congress, William Harper Davis spoke of "a program without parallel in history." "Never before," he added, "has there been a gathering of so large and representative a body of the world's leading scholars and thinkers."[1]

Administered by enterprising Americans like Howard J. Rogers, Nicholas Murray Butler, and Simon Newcomb, and financed by the federal government, the St. Louis Congress was indeed a momentous event. Over one hundred foreign scholars joined three times that number of Americans for a week's exchange of information in all fields of learning—from philosophy, anthropology, and psychology to medicine, law, and education. The official purpose of the Congress was to compensate for the increasing "subdivision and multiplication of specialties in science" by bringing "the scattered sciences into closer mutual relation."[2] One of the architects of the Congress, Harvard psychologist Hugo Münsterberg, sought to achieve this unification by organizing the fields of learning represented at the Congress into suggestive "divisions," "departments," and "sections." Although worried about being "too much influenced by the latest theories of to-day," Münsterberg divided the department on physics (within the division of physical science) into sections on matter, ether, and the electron.

He resisted current thought, however, by placing theoretical physics in the division of normative science alongside various fields of mathematics and philosophy, including methodology of science. To fill these various sections, the organizers of the Congress assembled an impressive group of international scientists including Ludwig Boltzmann, Henri Poincaré, Paul Langevin, Ernest Rutherford, and Wilhelm Ostwald. The joint meeting of the American Physical Society and the International Electrical Congress held during the previous week in St. Louis ensured a good turnout of American physicists.[3] What new messages were the Americans hearing in the speeches of the foreign delegates?

University of Paris professor Henri Poincaré (1854-1912) set the tone of the Congress with his speech on "mathematical physics"—a speech wherein he sought to evaluate contemporary physics by placing it within an historical context. In Poincaré's view, the nineteenth-century atomo-mechanical physics of "central forces" had given way, during the century, to the physics of "principles," such as the principle of conservation of energy. Perhaps, he speculated, the physics of "principles" was currently yielding to a yet unrevealed third phase of physics. Regarding the earlier "crisis" that signaled the transition from the physics of central forces to principles, Poincaré wrote: "Investigators gave up trying to penetrate into the detail of the structure of the universe, to isolate the pieces of this vast mechanism, to analyze one by one the forces which put them in motion, and were content to take as guides certain general principles which have precisely for their object the sparing us this minute study." These principles, which "are the results of experiments boldly generalized," included—aside from conservation of energy—the second law of thermodynamics, the equality of action and reaction, relativity of motion, conservation of mass, and the principle of least action. But Poincaré suspected that physics was "now upon the eve of a second crisis" and that all the trusted principles were "about to crumble away." He was concerned not only with the threat to the conservation of energy posed by radioactivity, "the grand revolutionist of the present time." Rather, he envisioned "a general ruin" of all the basic principles, save alone that of least action.[4]

The second law of thermodynamics, for example, was in danger. Recent interpretations of Brownian motion—the irregular move-

[143]

Ludwig Boltzmann *(left center)* awaiting St. Louis lecture by William Ostwald *(right center)*. From *Popular Science Monthly*, 1904. (Permission of Times Mirror Magazines, Inc.)

ment of small particles suspended in a fluid caused by molecular buffeting—supported the hitherto unverified view of the second law as an "imperfect" *statistical* theorem concerning molecular actions. The principle of the conservation of mass was also in question. Some physicists were proposing that there was no "true mechanical mass" but only mass of "exclusively electrodynamic origin." Many more were affirming that mass, whether mechanical or electrodynamic, varied with speed. These challenges to traditional thought suggested to Poincaré possible forms of the emerging, third phase of physics. Perhaps statistical mechanics "is about to undergo development and serve as model" for all of physics. "Then, the facts which first appeared to us as simple, thereafter will be merely results of a very great number of elementary facts which only the laws of chance make cooperate for a common end." Alternatively, with the demise of the principle of conservation of mass, perhaps there will soon emerge "a whole new mechanics . . . where inertia increasing with the velocity, the velocity of light would become an impossible limit."[5] This latter prediction, when added to an equally incisive evaluation of the principle of relativity,[6] has led historians to contend that Poincaré approached, in significant part, the breakthrough that Einstein actually achieved the following year, 1905. But it did not require latter-day historians to perceive the radical import of Poincaré's evaluation. In 1905, Arthur Lovejoy had already focused on Poincaré's "remarkable essay" and "the uncertainty and provisionality which are thus revealed in the foundation of the most fundamental of the physical sciences."[7]

Poincaré, while surveying recent transitions in the substantive content of physics, also traced the development of three distinct views of natural law. Prior to the adoption of the atomo-mechanical physics of central forces, scientists considered a law of nature to be "an internal harmony, static, so to say, and immutable; or it was like a model that nature constrained herself to imitate." With the advent of the physics of central forces, and the example of Newton's law of gravitation, scientists gradually discarded this "ancient," teleological view. Natural law became—and remained into the second and current phase of history, that of principles— simply "a constant relation between the phenomenon of to-day and that of to-morrow." In other words, natural law was pared down to a philosophically neutral and austere "differential equation." But what might be the fate of natural law in light of the current crisis in physics? Poincaré, envisioning a future physics based on statistical mechanics, hazarded a guess: "Physical law will then take an entirely new aspect; it will no longer be solely a differential equation, it will take the character of a statistical law." For Poincaré, in 1904, the ideology as well as the substance of nineteenth-century physics was exceedingly vulnerable to a "profound transformation."[8]

Equally convinced of fundamental changes occurring in physics was Paul Langevin (1872–1946), a young professor of physics at the Collège de France who had previously studied under J. J. Thomson and Pierre Curie. He felt that the scientific synthesis promised by the physics of the electron, his St. Louis topic, was like "a kind of New America, full of wealth yet unknown, where one can breathe freely, which invites all our activities, and which can teach many things to the Old World." Langevin's confidence in the "electronic conception of matter" rested on a double base. First, he was certain of electromagnetic theory (as developed by Faraday, Maxwell, and Hertz, and extended by Lorentz) and of an associated nonmaterial view of the ether; and secondly, he was sure that a recent group of independent experiments, the ones on charged water droplets by C. T. R. Wilson and J. J. Thomson, established the "granular structure of electricity." This two-fold confidence led him to advocate a complete inversion of traditional scientific thought. "It thus seems much more natural," he wrote, "to reverse the conception of Maxwell and to consider the analogy which he has pointed out between the equations of electromagnetism and those of dynamics under Lagrange's form as justifying much more the possibility of an electromagnetic representation of the principles and ideas of ordinary, material mechanics, than the inverse possibility."[9]

Of course, by 1904, this electromagnetic view was more than merely a possibility. In concluding, Langevin further emphasized: "Already all views, not only of the ether, but of matter, . . . obtain an immediate interpretation which mechanics is powerless to give, and this mechanics itself appears to-day as a first approximation . . . for which a more complete expression must be sought in the dynamics of the electron." In general, Langevin perceived the electromagnetic revolution to be well under way: "This idea has taken an immense development in the last few years, which causes it to break the framework of the old physics to pieces, and to overturn the established order of ideas and laws in order to branch out again in an organization which one forsees to be simple, harmonious, and fruitful."[10]

In early September of 1904, Ernest Rutherford of McGill University wrote to his close Yale colleague Bertram Boltwood, "I have been very busy preparing for the Scientific Congress at St. Louis and leave for St. Louis on Sept. 13." Rutherford (1871–1937) was working on a speech titled, "Present Problems of Radioactivity." In this lengthy but tightly knit talk, he appraised the latest experimental and theoretical knowledge regarding all phases of radioactivity from the nature of alpha, beta, and gamma radiations to the source of energy of radioactive bodies. Rutherford was especially concerned to show the amenability of radioactivity to Langevin's type of electromagnetic outlook. He was sympathetic to the idea that the mass of a beta particle, or electron, was entirely of electromagnetic origin and that the mass increased with speed up to the limiting speed of light. Moreover, he carefully pointed out that his and Frederick Soddy's radioactive disintegration theory, regarding questions of energy conservation, was in harmony with "the modern views of the electronic constitution of matter, which have been so ably developed by J. J. Thomson, Larmor, and Lorentz."[11]

Incidentally, although Rutherford was officially a "foreign delegate" to the St. Louis Congress, he had during the past few years become active in the United States physics community. A professorship from 1898 to 1907 at McGill University in Montreal enabled this New Zealander, educated at the Cavendish Laboratory, to participate regularly in American scientific gatherings held in Denver, Washington, St. Louis, and other cities. It was even possible for him to offer a summer course in 1906 at the University of California at Berkeley; and during the previous year, he had presented the Silliman Lectures at Yale. Arthur G. Webster, in his presidential address before the American Physical Society (APS) in February 1904, aptly characterized Rutherford's involvement in the American community: "It is a pleasure to recall that in the subject of radioactivity one of the leading contributors has been, if not an American, at any rate a professor in an institution on the American continent, a member of this society and of its council,

[147]

and a frequent contributor to its meetings."[12] Rutherford's participation in the APS insured that physicists in the United States kept abreast of the latest research in radioactivity. Michael Pupin of Columbia University went so far as to say that Rutherford's reports to the APS were themselves sufficient to justify the existence of the society. Thus, it is not surprising that Robert Millikan, an aspiring thirty-six-year-old professor at the University of Chicago, chose as the topic for a short paper at the St. Louis Congress, "The Relation Between the Radioactivity and the Uranium Content of Certain Minerals"—a topic also being studied by the scientist who was to become Rutherford's main American collaborator, Bertram Boltwood of Yale. It is also not surprising that in 1906 Yale joined other American universities in offering Rutherford a professorship. He declined all the offers, returning to England the following year.[13]

Wilhelm Ostwald (1853–1932), professor of physical chemistry at Leipzig, was the featured speaker on methodology and philosophy of science at the Congress. As William James pointed out three years later, Ostwald had for some time "been making perfectly distinct use of the principle of pragmatism in his lectures on the philosophy of science, though he had not called it by that name."[14] James could have been thinking of Ostwald's St. Louis lecture where he spoke about the necessity of establishing a definite and complete "correspondence" between the actual "manifold" of experience in a certain empirical situation, and the formal "manifold" of the scientists' conceptual constructions whether they be words, equations, or symbols. Regarding physical and chemical phenomena, Ostwald specifically insisted that the science of energetics, with its emphasis on measurable energy transformations, insured the fullest possible agreement between the empirical and conceptual manifolds. "One can perfectly characterize every physical event by indicating what amounts and kinds of energy have been present in it and into what energies they have been transformed. Accordingly, it is logical to designate the so-called physical phenomena as energetical." Moreover, he implied that in both physics and chemistry his newer, phenomenological, energetistic outlook was superior to the older, hypothetical, atomo-mechanical view. "If one applies to this question the criterion . . . given above, the completeness of the correspondence between the

[148]

representing manifold and that to be represented, there is no doubt that all previous systematizations in the form of hypotheses which have been tried in these sciences are defective in this respect."[15] On purely operational grounds—by requiring direct correspondence between concepts and empirical data—Ostwald found nineteenth-century mechanical physics to be lacking.

Although the operational and antiatomistic outlook of energeticist Ostwald was a long step from the early operational views of Charles Peirce and Simon Newcomb, it was an outlook with which Americans had become familiar well before the 1904 Congress. In particular, the newly founded *Physical Review,* under the editorship of Edward Nichols, regularly surveyed Ostwald's publications through the mid-to-late 1890s. The very first volume contained, for example, a discussion of Ostwald's two fundamental "laws of energetics" as presented in his 1893 *Textbook of General Chemistry.*[16] Edgar Buckingham, a Leipzig-trained physicist (Ph.D., 1893), proved to be a particularly staunch American advocate of antiatomistic energetics. Reporting in the *Physical Review* on a book by Georg Helm concerning chemical energetics, he wrote:

> At a time like the present, when both physics and chemistry are obscured by so many unnecessary hypotheses, every attempt to show the power and simplicity of general methods is very welcome. . . . The usefulness of thermodynamic, or energetic, methods has nowhere been greater than in the borderland between physics and chemistry. We are, however, inclined to think that pure physics may profit quite as much as physical chemistry from greater attention to the part which energy and its transformations play in natural phenomena, and that the greater advances in this direction made by physical chemistry are due rather to freedom from a long history of special hypotheses than to any intrinsic difference between the two fields.

Michael Pupin was another of Ostwald's American disciples even prior to the St. Louis lecture. Robert Millikan, one of Pupin's graduate students at Columbia from 1893 to 1895, later recalled that Pupin was "so much impressed" by energetics that he "did not believe in the kinetic theory at all."[17]

Not everyone in Ostwald's St. Louis audience, however, agreed with this energeticist's antiatomo-mechanical contentions. Milli-

[149]

kan afterward remembered that the "question of whether the atomic and kinetic theories were essential" was the Congress's "chief subject of debate." In fact, sitting in the front row attending Ostwald's lecture and joining in the following discussion was one of the oldest and most distinguished of the foreign representatives to the Congress, Ludwig Boltzmann (1840–1906), Ostwald's adversary from Leipzig and, more recently, Vienna.[18] On the day after Ostwald's lecture, Boltzmann made a formal rebuttal in his speech on theoretical physics delivered before the section on "Applied Mathematics." While granting that physics was presently in disarray—perhaps even "in process of revolution"—Boltzmann did not agree that generalized, phenomenological theories were better equipped to reestablish order than were specialized, atomo-mechanical hypotheses. Though appreciative of phenomenological theories, he denied that such theories were free from hypotheses or idealizations and hence irrefutable. "Without some departure, however slight, from direct observation," he maintained, "a theory or even an intelligibly connected practical description for predicting the facts of nature cannot exist." Moreover, Boltzmann felt that the usefulness of phenomenological theories was limited merely to summarizing or developing "knowledge previously acquired." On the other hand, specialized and admittedly tentative hypotheses "give the imagination room for play and by boldly going beyond the material at hand afford continual inspiration for new experiments, and are thus pathfinders for the most unexpected discoveries." Consequently, he rejected the accusation that the "development of mathematical methods for the computation of the hypothetical molecular motions has been useless and even harmful." He also felt that the recent web of experiments involving cathode rays and radioactivity added credence to the atomistic viewpoint. Boltzmann was particularly optimistic about statistical mechanics, especially Gibbs's formulation. Statistical mechanics, he concluded, would show scientists "the facts of experience in an entirely new light" and would consequently inspire them to "new thought and reflection."[19] In retrospect, we realize that this re-thinking had already begun in 1900; quantum theory was taking shape following Max Planck's application of Boltzmann's statistical techniques to radiating bodies.

[150]

Two prominent European physicists absent from the Congress—J. J. Thomson and H. A. Lorentz—also visited the United States around 1904. In his Silliman Lectures at Yale in 1903, Thomson (1856-1940) brought his American audience up to date on the latest developments concerning "Electricity and Matter," including his so-called "plum-pudding" electron model of the atom. A few years earlier, in 1896, Thomson, the director of the Cavendish Laboratory, had also given four lectures at Princeton on electrical discharges in gases, the photoelectric effect, and cathode rays interpreted as subatomic electrical corpuscles.[20] Lorentz (1853-1928) spent two months in 1906 at Columbia University presenting a comprehensive course of lectures on the subject which he had pioneered: "The Theory of Electrons and Its Applications to the Phenomena of Light and Radiant Heat." Interestingly, as late as 1909, when he amended and published these lectures, this sophisticated Leiden physicist, like many of his colleagues around the world, was only beginning to grasp the full importance of Planck's quantum theory let alone Einstein's more recent theory of relativity. In very brief passages, Lorentz only tentatively endorsed these two radical perspectives, preferring, in the case of relativistic phenomena, his own electromagnetic interpretation.[21]

Ernst Mach (1838-1916), another prominent European physicist and philosopher who was absent from the Congress, also came to the United States—not in person, but by way of the printed word. Thanks to a dedicated American publisher, the Open Court Company of Illinois, physicists in the United States became acquainted with Mach's antimetaphysical critique of atoms and classical mechanics as well as his positivist belief in "economy of thought." Under the direction of German-born editor Paul Carus, and using translations mainly by assistant editor Thomas McCormack, the company's journal, *Open Court,* printed numerous articles by Mach in the years around 1900. More important, the Open Court Company published various English editions of Mach's *Science of Mechanics, Popular Scientific Lectures,* and *Analysis of the Sensations.* By the way, McCormack also used the pages of *Open Court* to enthusiastically endorse Stallo's recently reissued *Concepts and Theories of Modern Physics.* Noticing the similarities between the views of Stallo and Mach, he specifically praised Stallo for his early "grasp of what are now acknowledged principles of scientific criticism. . . ."[22]

Critics for the *Physical Review* generally applauded both the first (1893) and second (1902) American editions of Mach's *Science of Mechanics.* (Charles Peirce had helped prepare the first edition.) Particularly pleased that there had been sufficient demand for the second edition, Edward Nichols recommended that it be placed "in the hands of every teacher of physics." If Open Court's "Testimonials of Prominent Educators" are to be trusted, Michael Pupin of Columbia, William Magie of Princeton, and Henry Crew of Northwestern had already agreed, by 1897, with Nichols's favorable recommendation regarding Mach's iconoclastic *Science of Mechanics.*[23] Mach's influence was so pervasive that it even reached the secondary schools. During his senior year at high school (1899–1900), Percy Bridgman read Mach's writings as well a related books on the conceptual foundations of science by Karl Pearson, William Clifford, and Stallo. And by about 1910, American educators were routinely recommending that high-school teachers familiarize themselves with Mach's works along with those of Pearson, Poincaré, Ostwald, Clifford, and Stallo.[24]

15

AMERICANS AT ST. LOUIS

American physicists at the St. Louis Congress agreed with what Poincaré, Langevin, Rutherford, Ostwald, and Boltzmann were saying: that physical science was experiencing a dramatic upheaval. The Americans, while less personally involved in this restructuring of science than their foreign colleagues, were well-informed about the latest electromagnetic, thermodynamic, and statistical-mechanical alternatives to the faltering atomo-mechanical physics. And, although less articulate than their overseas visitors, they generally advocated the newer phenomenological, operational, or positivistic attitudes toward scientific inquiry and physical law. Indeed, the Americans revealed their receptivity to modern ideas merely by inviting to the Congress such an innovative group of foreign representatives.

Who were these progressive members of the American delegation? For the most part, they were a generation younger than Mayer, Gibbs, Trowbridge, and Rowland; they were born in the 1850s and 1860s. They were also more geographically dispersed than their predecessors, who were concentrated in a few universities along the east coast. In addition, they were leaders in the burgeoning American physics community, serving as either officers or fellows in the American Association for the Advancement of Science or the recently founded American Physical Society. And all had completed doctoral degrees, many having studied either at Johns Hopkins or in Germany or both. They included: Carl Barus (1856–1935), physicist at Brown University with an earlier background in geophysics; Arthur Kimball (1856–1922), an Amherst professor; Dewitt Brace (1859–1905), a University of Nebraska researcher with a special interest in the ether; Henry Crew (1859–1953), a professor at Northwestern; and the aforementioned Robert Millikan (1868–1953), a physicist at the University of Chicago. A slightly more moderate member of this group was Edward Nichols (1854–1937), who was a Cornell professor and cofounder, in 1893, of the *Physical Review*. [1] Nichols at age fifty was probably

[153]

Robert Millikan. (J. Hagemeyer, Bancroft Library, University of California, Berkeley; courtesy of AIP Niels Bohr Library)

the most prominent American physicist at the Congress: Rowland had died in 1901, Gibbs in 1903, and Michelson was absent from the St. Louis meetings because of scheduling and minor health problems.[2]

Of course, not all the Americans were liberal or moderate, let alone young and German-educated. As mentioned earlier, the president of the Congress and keynote speaker was sixty-nine-year-old Simon Newcomb, a man most at home with the science of the nineteenth century. Similarly, Francis Nipher, although only fifty-seven, spoke out at the Congress in favor of threatened mechanical positions. And there were other conservatives. Robert Woodward (1849–1924), president of the newly founded Carnegie Institution of Washington, opened the meetings of the Division of Physical Science with remarks true to the scientific ethos of the recent past. In words reminiscent of Trowbridge in the 1880s, he hoped for a grand atomo-mechanical unification of all physical phenomena: "the day seems not far distant when there will be room for a new *Principia* and for a treatise which will accomplish for molecular systems what the *Mécanique Céleste* accomplished for the solar system."[3] Our concern in these final pages is no longer with traditionalists like Nipher and Woodward, but with progressives like Barus, Kimball, Brace, Crew, and Nichols.

The electromagnetic view of nature was the dominant new outlook among the Americans in St. Louis. Carl Barus capped his encyclopedic survey of "The Progress of Physics in the Nineteenth Century" by describing the recent and "splendid triumph of the electronic theory." In his opinion, the ether-vortex hypothesis of previous decades had proved incapable of explaining the mass of an atom or the principles of dynamics; the electronic theory of Lorentz, J. J. Thomson, Joseph Larmor, Max Abraham, and others held greater promise. Barus maintained that "not only does this new electronic tendency in physics give an acceptable account of heat, light, the x-ray, etc., but of the Langrangian function and of Newton's laws."[4] Though admitting to have been "startled at first by the very audacity of this theory," Arthur Kimball echoed Barus's sentiment. Kimball was particularly impressed by J. J. Thomson's "most remarkable" theory of the atom as a swirl of hundreds of electrons. In general, he concurred that physicists should "seek the explanation of matter and its laws in terms of

Henry Crew, 1933. (Northwestern University Archives, courtesy of AIP
Niels Bohr Library)

the properties of ether and electricity, instead of trying to unravel the secrets of electricity and ether in terms of matter and motion."[5]

Dewitt Brace agreed. An internationally respected expert on the ether, Brace reported to the Congress that he was experimentally testing the implications of electron theory regarding, for example, mass variation and length contraction.[6] Finally, while Edward Nichols accepted the unprecedented empirical effects associated with electron theory, he preferred to interpret the effects in terms of an underlying, abstract ether rather than in terms of primitive, "disembodied" electric charge. In the increasingly familiar statement, "Matter is composed of electricity and of nothing else," he preferred to substitute for the hypothetical concept of "electricity" the equally "imaginary" but more tractable concept of "ether." "If matter is to be regarded as a product of certain operations performed upon the ether," he concluded, "there is no theoretical difficulty about transmutation of elements, variation of mass, or even the complete disappearance or creation of matter."[7]

These Americans were not naive champions of the electron theory and related views. Along with their European colleagues, they realized that they were dealing with a nascent theory, incomplete and unproven. Indeed, so penetrating were some of their criticisms that physicists would resolve them only after accepting and developing Planck's quantum hypothesis and Einstein's theory of relativity. Kimball pinpointed the tasks still to be accomplished by J. J. Thomson's electron model of the atom. He could only hope that Thomson's model would eventually account for specific heats of gases—that persistent problem recently resurrected by Lord Kelvin as one of the "nineteenth-century clouds over the dynamical theory of heat." Similarly, he could only trust that the model would someday explain complex atomic spectra. And he worried that the model still lacked an agency for assigning distinct numbers of electrons to the atoms of particular chemical elements. "Some kind of natural selection seems to be needed," Kimball wrote, "to explain why some atoms having special numbers of corpuscles survive while intermediate ones are eliminated." Brace also had reservations about the electron theory. He was dissatisfied with the ad hoc character of recent electromagnetic interpretations of ether-drift experiments—interpretations that were "highly artificial" in their "successive auxiliary hypotheses and approximations."[8] In

Edward Nichols. (Cornell University Archives, courtesy of AIP Niels
Bohr Library)

sum, American physicists understood and appreciated the electromagnetic view, but they neither extolled nor defended it dogmatically.

The Americans at the Congress coupled their conditional endorsement of the electron theory with an open wariness of prior mechanical outlooks. Barus stressed that all physicists now accepted Maxwell's electromagnetic equations as accurate descriptions of phenomena, but most rejected the atomo-mechanical "methods by which Maxwell arrived at his great discoveries." Along the same line, in a paraphrase of Poincaré, Barus emphasized the arbitrariness of all ethereal or material interpretations of electromagnetism: "If, says Poincaré, compatibly with the principle of conservation of energy and of least action, any single ether mechanism is possible, there must at the same time be an infinity of others." Kimball also called attention to the heuristic nature of the mechanical postulates traditionally associated with theories of matter, such as the vortex-atom theory. "It is entirely natural that such ideas as impenetrability and inertia, borne in upon us as they are by our experience of matter in bulk, should affect our theorizing," Kimball acknowledged, "but it should never be forgotten that as fundamental postulates they have no more authority than any others that might be assumed that will coordinate the same facts of observation."[9] Like many of their contemporaries, Barus and Kimball considered mechanical models and concepts to be manifestly hypothetical.

Not every American at the Congress aligned himself with the electromagnetic alternative to traditional mechanical physics. Henry Crew opted for a phenomenological approach based on a few broad, empirical principles. Though chairman at the Congress of the Department of Physics and of the Section on Physics of Ether, Crew did not deliver a formal lecture. Earlier in 1904, he had outlined his general thoughts in a speech later published as the opening article in *Science*. Speaking on "Recent Advances in the Teaching of Physics," Crew espoused educational goals that reflected two related trends: skepticism of atomo-mechanical speculations and acknowledgement of the limits of scientific inquiry. He insisted: "The modern instructor does not trifle with atoms, molecules and other hypothetical creatures which he has not seen and does not know about. He takes pains to point out the line of demarcation between the known and the unknown, believing that few

[159]

things are more instructive for the learner than the limitations of human knowledge concerning even household matters." Crew, who in later years became president of the History of Science Society (1930) as well as of the American Physical Society (1909), accented this latter point with a personal example from the recent history of the teaching of physics: "As a boy I was taught to respect Newton as the man who had explained gravitation; to-day the lad is taught that Newton distinctly refused even to make a guess at its explanation."[10]

In place of atomo-mechanical doctrines learned solely through textbooks, Crew advocated the "energy viewpoint" learned through first-hand student experiments. As early as 1900, he had elaborated on this same theme, arguing that "dependence upon the words of the text, and reckless ideas concerning molecules and atoms are, I fear, responsible for much of the learning that has to be unlearned." Like Poincaré and Ostwald, he had called for the unification and simplification of physics through "perhaps four or five, general experimental principles." The principles included conservation of energy, conservation of matter, least action, Newton's laws of motion, and the general laws of wave motion. Upset "with vague notions concerning *force* and *energy,* with loose definitions of *molecules* and *atoms*—definitions, indeed, that do not define," he had contended that the general principles must "be experimentally demonstrated in such a way as to leave no doubt in the mind of the student regarding either their meaning or their validity."[11] As envisioned by Crew, a phenomenalistic emphasis on general principles went hand in hand with the relatively new and enlightened methods of laboratory teaching.

Edward Nichols, although more disposed toward traditional mechanical precepts than Crew, also regarded these precepts with a discriminating, somewhat positivistic attitude. In his St. Louis speech, he explained and defended the idea of evaluating scientific concepts through the use of the "dimensional formula"—a familiar idea traceable to Fourier, Maxwell, and others, but given systematic presentation only around 1920 by an international group that included Edgar Buckingham and Percy Bridgman.[12] Nichols believed that any physical quantity should ultimately be expressible as a relationship between three empirically certain, mechanical concepts: mass (M), distance (L), and time (T). Thus, the dimensional formula for acceleration was LT^{-2}; for force $LT^{-2}M$; and for energy L^2MT^{-2}. Nichols was enthralled by this mode of analysis primarily because it "affords a valuable criterion of the extent and boundaries of our strictly definite knowledge of physics. Within these boundaries we are on safe and easy ground, and are dealing, independent of all speculation, with the relation between precisely defined quantities." That is, if a physicist could assign a definite dimensional formula, free of factors other than M, L, and T, to a particular quantity such as force or energy, then he was dealing with "that positive portion of physics, the mechanical basis of which, being established upon direct observation, is fixed and definite, and in which the relations are absolute and certain as those of mathematics itself."[13]

On the other hand, if a physicist was unable to assign a pure dimensional formula to a quantity, then he was dealing with an "incomplete" and "speculative" realm of science. Nichols maintained that such was currently the case with studies on the microscopic workings of heat, light, and electromagnetism; it was also the case with researches on the ultimate properties of ether and matter. In general, like positivist Ernst Mach, Nichols was content to view science as "nothing more than *an attempt to classify and correlate our sensations.*" Thus, he stated that force, expressible in a dimensional formula as $LT^{-2}M$, "is simply a quantity of which we need to know only its magnitude, direction, point of application, and the time during which it is applied. . . . All troublesome questions as to how force acts, of the mechanism by means of which its effects are produced, are held in abeyance."[14]

Though his overall outlook contained vestiges of mechanical reductionism, Nichols presented at St. Louis a perceptive appraisal of current physics. He sharply distinguished "positive" and "speculative" knowledge and forcefully called for precise definitions built on direct observation. He thus joined delegates Barus, Kimball, and Crew as an American spokesman for the progressive currents in turn-of-the-century physics.

[161]

16

FRANKLIN'S OPERATIONAL PERSPECTIVE

William Franklin, a former student of Edward Nichols, went a step further than most American physicists who were grappling with the diverse and controversial strands of contemporary science. At the beginning of the new century, he enunciated a generalized, all-embracing operational perspective on the conceptual foundations of physics.

Franklin (1863–1930) was both a product and a respected member of the American physics community. The physicist who most influenced him during his student years was Nichols, a man well informed, as we have just seen, about turn-of-the-century scientific options, including Machian positivism. Franklin studied under Nichols at two different universities: he received his bachelor's and master's degrees in 1887 and 1888 at the University of Kansas; thirteen years later, in 1901, he was awarded a doctorate at Cornell University. During this period, he collaborated with Nichols in writing various technical articles and a three-volume textbook. In the early 1890s, he also studied at the University of Berlin and at Harvard University, spending a year at each institution. If we judge by the frequency with which Franklin later quoted his Berlin professor, Hermann von Helmholtz, it appears that Helmholtz shared with Nichols the distinction of being a key influence on Franklin. Specifically, he helped mold Franklin's antipathy toward metaphysical speculation but tolerance toward mechanical models when treated merely as useful conceptual constructions. Franklin's first major teaching post was at Lehigh University, where he was professor of physics and electrical engineering from 1897 to 1903 and later, professor of physics from 1903 through 1915. He next accepted a position as professor of physics at the Massachusetts Institute of Technology, serving from 1918 until his retirement in 1929. During these later decades, he also gave courses at Columbia and Harvard. Although Franklin never excelled as a researcher, he did achieve national prominence as an educator in both physics and electrical engineering. The author or coauthor of over two dozen textbooks and numerous general essays, he posthumously received the first teaching award of the American Association of Physics Teachers.[1]

As early as 1902, Franklin advocated an operational perspective on contemporary physics. It was in that year that the thirty-nine-year-old Lehigh professor addressed the AAAS as vice-president in charge of physics. One of his central themes was that this science should be taught as a body of conceptual constructions that are definable only through actual operations. Specifically, he held that "the characteristic feature of the study of science is a *determining objective constraint upon the processes of the mind."* Though an abstruse pronouncement, his meaning became clear with an example:

> What is the definition of the mass of a body? The careless and imaginative definition which is usually given is that "the mass of a body is the quantity of matter the body contains." I suppose that definition satisfies many of you, but it does not satisfy me. All our notions of length and angle take their rise in and are fixed or defined by those fundamental geometric operations of congruence. The real definition of mass is no less a physical operation, the verbal definition is the briefest possible specification of this operation and it can be nothing else, the result of this operation on a given body is an invariant number, and by a feat of the imagination we conceive this invariant number to be a measure of the quantity of matter the body contains.[2]

It is perhaps surprising that Franklin's consistent use of the term "operation" in this and subsequent passages antedated, by over two decades, Percy Bridgman's publication of *The Logic of Modern Physics,* the book usually deemed to be the first as well as the definitive statement of the "operational attitude." We have seen in earlier chapters, however, that the general idea behind Franklin's use of the term was neither new nor unusual around 1900. This was not only a period well after the initial pronouncements of Wright, Newcomb, and Peirce but also a period of activity for Mach, Poincaré, Pearson, and Ostwald. It was also a time of renewed appreciation for Stallo and Clifford. In addition, the turn of the century marked the popular appearance of pragmatism as variously construed by Peirce, William James, and John Dewey. Even Edwin Hall, a physicist with an aversion to philosophy, at-

[163]

William Franklin. (The MIT Museum and Historical Collections)

tempted during these years to understand what both Peirce and James meant by pragmatism.[3]

Franklin proceeded in his AAAS address to criticize educated laymen along with his fellow scientists for occasionally forgetting that most definitions in physics were "in reality operations." That is, persons tended to forget that physical concepts are devoid of meaning when considered independently of their associated, defining operations. The consequence was a tendency "to confuse the boundaries between our logical constructions and the objective realms which they represent to the understanding." Although scientists habitually spoke of concepts "in objective terms for the sake of concreteness and clearness," he insisted that these concepts were "purely notational" abstractions. In particular, atomistic concepts, while useful mental tools, were "mere logical constructions."[4] Moreover, as he spelled out a few years earlier in a review of a new treatise on thermodynamics by Pierre Duhem, even the thermodynamic—and hence purportedly phenomenalistic—explanations of energeticists often involved "conceptual" or, more accurately, "algebraic constructions." "Very many of the most striking and useful developments of thermodynamics," he elaborated, "are the results of special hypotheses of an unreal character incapable in themselves of experimental verification." For example, "energy—the kind we think about—is nothing more than a generalized notion not to be properly thought of as existing outside of one's head except for the sake of concreteness."[5]

Franklin concluded his address with a challenge to his audience of physicists, most of whom were also professors and teachers:

> I wonder if any of you really doubt that every notion in physics, definite or indefinite, is associated with and derived from a physical operation, and that absolutely the only way to teach physics to young men is to direct their attention to that marvelous series of determining operations which bring to light those one-to-one correspondences which constitute the abstract facts of physical science. If you do [doubt this], I am bound to say I do not think much of your knowledge or teaching of physics.

With equal candor, he went on in two final sentences to capture the prevailing sentiments against hypothetical atomo-mechanical attitudes and in favor of cautious operational outlooks: "I think

[165]

that the sickliest notion of physics, even if a student gets it, is that it is 'the science of masses, molecules and the ether.' And I think that the healthiest notion, even if a student does not wholly get it, is that physics is the science of the ways of taking hold of bodies and pushing them!'"[6] For Franklin, as for many of his American colleagues in the turbulent years around 1900, all pretensions of penetrating the ultimate secrets of the physical universe had vanished. Whether pertaining to atoms, ether, or energy, a physical concept was now acknowledged to be a conceptual construction, meaningful only to the extent that it could be uniquely associated with actual concrete operations.

Franklin continued to advocate the operational view throughout his career: for example, in a 1918 textbook, he introduced a discussion of heat with a brief essay titled "Operative versus Inoperative Definitions"; and in June 1926, the lead article in *Science* was his "Operative Versus Abstract Philosophy in Physics."[7] During these later decades, his conviction that physical concepts were "mere logical constructions" very likely correlated with his receptivity to the new relativity and quantum theories. Because he avoided treating any concepts as "objective realities," he was open to the possibility of alternative conceptual constructions, as long as they met the test of being operationally definable. In 1911, only two years after the appearance of the first American article on relativity by Gilbert Lewis and Richard Tolman, Franklin published a clear exposition and endorsement of Lorentz and Einstein's "Principle of Relativity." And by the mid-1920s, he was presenting explications of Niels Bohr's quantum theory of the atom. With typical openmindedness and enthusiasm, he predicted in a 1924 article in *Science* that the new quantum theory would soon lead to even "a more wonderful transformation of our conceptions of the physical world . . . than that which has resulted from the relativity theory." When this transformation actually came a few years later with the matrix and wave mechanics of Werner Heisenberg and Erwin Schrödinger, Franklin persisted in keeping abreast of the dramatic developments, even after his retirement from MIT in 1929.[8]

[166]

17

ADAMS: FROM UNITY TO MULTIPLICITY

Atomo-mechanical outlooks, although dominant, were neither mono-lithic nor unchallenged in the United States from about 1870 to 1895. We have seen that they were diversified and in flux, and that they existed alongside a number of nascent alternatives. But the next ten-year period, the decade roughly from 1895 to 1905, was, for American physics, a time of even greater intellectual ferment; it was then that a quickening of prior movements occurred. Alert to changes in the realms of experiment, theory, and scientific ideology which were taking place internationally, Americans increasingly embraced the newest trends in physics. Henry Adams, in that rich and imagina-tive autobiographical account of his life and times, *The Education of Henry Adams,* presented a perceptive, contemporary evaluation of the latter upheaval. It seems appropriate to conclude our study with the views of this nonscientist on the state of physics around 1900; recall, if you will, that we began with the observations of another perceptive nonprofessional, John Stallo, who addressed himself to the same topic over twenty years earlier.

Adams (1838–1918), a Harvard alumnus and descendant of two American presidents, was by century's end an accomplished and respected historian, author, and teacher. In the *Education,* written for the most part in 1905 at age sixty-seven, he recounted the efforts he made during the preceding few years to try to understand recent trends in physics.[1] He hoped through the study of modern physics not only to further his life-long quest for "education," but also to acquire insights applicable to his scientific theories of history. With this latter goal in mind, he sought in the late 1890s an informal tutor in physics. Simon Newcomb, one of the "most distinguished" American physical scientists, was uninterested in Adams's type of scientific-historical aspirations, while J. Willard Gibbs—"the greatest of Americans, judged by his rank in science"—did not live in Wash-ington, Adams's home city. Consequently, he turned to Samuel Langley, a distant relative and close friend who was the director of the Smithsonian. Langley, who in Adams's words "nourished a

Henry Adams, 1872. (Courtesy of Harvard University Archives)

scientific passion for doubt," oversaw much of Adams's initiation into physics.[2] To Langley's tutoring, Adams added extensive reading in the most up-to-date works of world leaders in physical science. Detailed summaries in the *Education* as well as discerning marginal comments in his personal copies of the scientific volumes attest to his immersion in the works of Henri Poincaré, Wilhelm Ostwald, and Ernst Mach.[3]

At times writing with childlike enthusiasm, at other times with sarcastic dissatisfaction, Adams reported in the *Education* that a "cataclysm" of unprecedented magnitude occurred in physics around 1900. Prior to this, physicists were blissfully building their science around a cluster of lofty and transcendent concepts: "Unity, Continuity, Purpose, Order, Law, Truth, the Universe, God." Newton, Maxwell, and other former physicists were tackling the unknowns of the physical world confident that they were confronting, as Tennyson expressed it:

> One God, one Law, one Element,
> And one far-off, divine event,
> To which the whole creation moves.

But then, Adams observed, "suddenly, in 1900, science raised its head and denied." Breaking radically with the attitudes of past science—indeed, of past western culture—physicists began to speak of "Multiplicity, Diversity, Complexity, Anarchy, Chaos." "The child born in 1900 would, then," Adams summarized, "be born into a new world which would not be a unity but a multiple."[4]

Adams drew much of this interpretation not only from the works of Poincaré, Mach, and Ostwald, but also from the second edition (1900) of Karl Pearson's popular *Grammar of Science*. Pearson (1857–1936), a British scientist and statistician, aimed in his book to evaluate the foundations or "grammar" as opposed to the superstructure or everyday "language" of science, particularly physics.[5] In a chapter of the *Education* appropriately titled "Grammar of Science," Adams reported Pearson's basic displeasure with Lord Kelvin's traditional mechanical interpretation of force and matter. In particular, Adams focused on Pearson's staunch sensationism: "Pearson shut out of science everything which the nineteenth century had brought in. He told his scholars that they must put up with a fraction of the universe, and a very small fraction at that—the circle reached by the senses, where sequence could be taken for granted." Pearson objected not only to the substantive content of nineteenth-century mechanical physics, but also to the related ideology, which he found both metaphysical and materialistic. To convey this point, Adams quoted Pearson verbatim:

> Order and reason, beauty and benevolence, are characteristics and conceptions which we find solely associated with the mind of man. . . . Into the chaos beyond sense-impressions we cannot scientifically project them. . . . In the chaos behind sensations, in the 'beyond' of sense-impressions, we cannot infer necessity, order or routine, for these are concepts formed by the mind of man on the side of sense-impressions. . . . Briefly, chaos is all that science can logically assert of the supersensuous.

"In plain words," Adams concluded, "Chaos was the law of nature; Order was the dream of man."[6]

Adams noticed a number of similarities between Pearson's *Grammar* and an older volume to which Langley had directed him in the late 1890s, Stallo's *Concepts and Theories of Modern Physics*. But there was one revealing difference. On the one hand, scientists had warmly received the *Grammar*. (To support this observation, Adams could have cited, for example, the reviewer in *Science* who argued that the *Grammar* was "creditably and suggestively" aiding professional scientists.[7]) On the other hand, Adams was aware that scientists had greeted Stallo's *Concepts and Theories* with disdain. In these contrasting receptions, he found a vivid demonstration of the general shift that had occurred in scientific outlooks: "The progress of science was measured by the success of the 'Grammar,' when for twenty years past, Stallo had been deliberately ignored under the usual conspiracy of silence inevitable to all thought which demands new thought-machinery. Science needs time to reconstruct its instruments, to follow a revolution in space; a certain lag is inevitable; the most active mind cannot instantly swerve from its path."[8] Although it was an overstatement to attribute complete novelty of thought to Stallo and to accuse scientists of conspiring to ignore that thought, Adams's general point regarding the contrasting receptions in differing conceptual climates was valid.

[170]

It was also during 1905, in an introduction to the American edition of Poincaré's *Science and Hypothesis,* that Harvard philosopher Josiah Royce contrasted the reception given Stallo's book to the recognition given current works of thinkers like Poincaré. He remembered that, following the publication of Stallo's book, the "sense of scientific orthodoxy was shocked amongst many of our American readers and teachers of science"; Royce continued:

> That very book, however, has quite lately been translated into German as a valuable contribution to some of the most recent efforts to reconstitute a modern 'philosophy of nature.' And whatever may be otherwise thought of Stallo's critical methods, or of his results, there can be no doubt that, at the present moment, if his book were to appear for the first time, nobody would attempt to discredit the work merely on account of its disposition to be agnostic regarding the objective reality of the concepts of the kinetic theory of gases, or on account of its call for a logical rearrangement of the fundamental concepts of the theory of energy. We are no longer able so easily to know heretics at first sight.[9]

Like Adams, Royce perceived that Stallo's antimetaphysical critique of atomo-mechanical physics was now nearer the norm of scientific thought. This was especially apparent since 1901 when the book received the sanction of a German translation, instigated by no less a figure than Ernst Mach.[10]

Even though Adams maintained that Stallo was completely ignored for two decades and that a "cataclysm" in physics occurred around 1900, he realized on closer view that the transition in thought had not been so abrupt. Adams himself "was prepared to expect" the transition "under the silent influence of Langley." During past years, he had followed Langley's Smithsonian *Annual Reports* in which "the revolutionary papers that foretold the overthrow of nineteenth-century dogma" were reprinted. Adams remarked that papers by Crookes, Roentgen, and Curie "had steadily driven the scientific lawgivers of Unity into the open." More generally, he believed that any person reflecting on the discovery of X rays and radioactivity would appreciate that the change, though dramatic during the short period of time since 1900, "had not been so sudden as it seemed." The "flow of the tide" had been perceptible for at least twenty years, until it had "become marked"

[171]

in 1895 with Roentgen's observation of X rays and, in the next few years, with Becquerel's discovery of radioactivity and the Curies' isolation of radium and polonium. "The man of science must have been sleepy indeed," Adams added, "who did not jump from his chair like a scared dog when . . . Mme. Curie threw on his desk the metaphysical bomb she called radium."[11]

One of Adams's favorite metaphors for late nineteenth-century physics was a ship wrecked at sea. Alfred Mayer had toyed with a similar metaphor in his 1882 review of Stallo's *Concepts and Theories of Modern Physics.* Confronted by Stallo's assault on bulwarks of physics like the undulatory theory of light and the kinetic theory of gases, Mayer had asked with rhetorical disbelief: "Is it expected that physicists will at once abandon the carefully-framed vessels on which they have sailed to discoveries so triumphant, with not even a raft prepared on which to take their chances?" Twenty-three years later, Adams reported that physicists not only had abandoned the now "wrecked" vessels of Mayer's day but also had resigned themselves to an austere, newly fashioned "raft." The continually quickening storms of recent decades, Adams concluded, had left "science adrift on a sensual raft in the midst of a supersensual chaos."[12]

In Adams's America, the unity hoped for but never achieved in nineteenth-century physics had given way to the multiplicity of the emerging but as yet unsettled scientific patterns of the new century. Influenced by and alongside their European colleagues, Americans had gradually abandoned efforts to explain all physical phenomena in terms of the hypothetical atoms and ethers of classical mechanics; instead, they were turning to a diversity of electromagnetic, thermodynamic, and statistical-mechanical outlooks on nature. And in place of the traditional goals of seeking ultimate truths and absolute natural laws, Americans increasingly were resigning themselves to philosophically agnostic, mathematically abstract, operational accounts of physical phenomena. All this was occurring before the impact of Planck and Einstein. The stage was thus being set in the United States for the advent of the new physics.

NOTES

INTRODUCTION

1. Harman, *Energy, Force, and Matter,* pp. 154–55.
2. Adams, *The Education of Henry Adams,* ed. Samuels, pp. 449–61. (Discussed later, chap. 17.)
3. Kevles, "The Physics, Mathematics, and Chemistry Communities," in *The Organization of Knowledge in Modern America,* eds., Oleson and Voss, pp. 139–72.
4. Forman, Heilbron, and Weart, "Physics circa 1900," p. 5.
5. Sopka, *Quantum Physics in America.* Goldberg, "Early Response to Einstein's Theory of Relativity." Badash, *Radioactivity in America.* Kevles, *The Physicists.* For the quotations from Heisenberg and Einstein, see Holton, "The Migration of the Physicists to the United States," forthcoming in an Einstein centennial volume published by the Smithsonian Institution.

1. STALLO'S CRITIQUE

1. Stallo, *The Concepts and Theories of Modern Physics,* pp. 7–9, 15, 17–18, 23, 27. The first edition of this book was also published in 1882 by D. Appleton and Co. in New York; however, both the copyright and Stallo's preface were dated 1881.
2. For Stallo's biography, see Stallo, *The Concepts and Theories of Modern Physics,* 3d ed., ed. Bridgman, pp. x–xvi. See also Drake, "J. B. Stallo and the Critique of Classical Physics," pp. 32–37.
3. Stallo's essays, including "Materialismus," were collected and reprinted as *Reden, Abhandlungen und Briefe.* See also Easton, *Hegel's First American Followers,* pp. 32–57.
4. Leverette, "Science and Values," p. 48 and chap. II. Walker, "The Popular Science Monthly, 1872–1878," pp. 39–40.
5. Stallo, "The Primary Concepts of Modern Physical Science," *Pop. Sci. Mo.,* 3 (1873), pp. 705–17, and 4 (1873–74), pp. 92–108, 219–31, and 349–61. For comments on Tyndall, see in particular 4 (1873–74), pp. 92, 219; for Du Bois-Reymond, see 3 (1873), p. 705 and 4 (1873–74), pp. 358–59.
6. Youmans, "Editor's Table," *Pop. Sci. Mo.,* 3 (1873), pp. 771–72; Harris, "Notes and Discussion," *Journal of Speculative Philosophy,* 7 (1873), p. 90. The February 1874 issue of the *Pop. Sci. Mo.,* however, did print two letters that were critical of certain obscure points in Stallo's articles; see "Correspondence," *Pop. Sci. Mo.,* 4 (1873–74), pp. 493–95.
7. Stallo, *Concepts and Theories,* pp. 7, 24, 83, 129, 294. For a study of Stallo's "theory of cognition," see Wilkinson, "John B. Stallo's Criticism of Physical Science." See also *Microfilm Abstracts,* 11 (1951), pp. 496–97.
8. Stallo, *Concepts and Theories,* pp. 24, 27–29.

[173]

9. In this paragraph, I follow Stallo's argument in Chap. III of *Concepts and Theories*; see esp. pp. 30–31, 33–38.

10. In this and the next paragraph, I follow Stallo's argument in Chap. IV of *Concepts and Theories*; see esp. pp. 40–44.

11. Stallo, *Concepts and Theories*, p. 29, 61. In this paragraph, I follow Stallo's argument in Chap. V; see esp. pp. 52, 55–56, 58–65.

12. Stallo, *Concepts and Theories*, p. 29. In this paragraph, I follow Stallo's argument in Chap. VI; see esp. pp. 66–68, 82. See also Moyer, "Energy, Dynamics, Hidden Machinery," pp. 251–68.

13. In this paragraph, I follow Stallo's argument in Chap. VII of *Concepts and Theories*; see esp. pp. 84–86, 92–99.

14. In this paragraph, I follow Stallo's presentation in the first half of Chap. VIII of *Concepts and Theories*, pp. 104–17.

15. Stallo, *Concepts and Theories*, p. 101.

16. Stallo, "The Primary Concepts of Modern Physical Science: Part II— The Atomic Constitution of Matter as a Postulate of Thought," *Pop. Sci. Mo.*, 4 (1873–74), p. 94. See also Cantor and Hodge, *Conceptions of Ether*, pp. 1–60.

17. In this and the next paragraph, I follow Stallo's argument in the second half of Chap. VII of *Concepts and Theories*, pp. 119–28. See also Brush, *The Kind of Motion We Call Heat*, pp. 63–65.

18. Stallo, *Concepts and Theories*, pp. 137, 148–51, 183–87.

19. Ibid., p. 294.

2. REACTIONS OF REVIEWERS

1. Royce's comment appears in his introduction to Poincaré, *Science and Hypothesis*, p. xvi.

2. Leverette, "Science and Values," pp. 33–34; Walker, "The Popular Science Monthly," pp. 69–74.

3. "Stallo's Theories and Concepts of Modern Physics," *The American*, 3 (11 Feb. 1882), pp. 281–82. The author of this unsigned review is identified as "C.D. English" in *Poole's Index to Periodical Literature*, II, p. 341. "Concepts and Theories of Modern Physics," *The Literary World*, 13 (11 March 1882), p. 72. "The Atomo-Mechanical Theory," *The American Engineer*, 4 (21 July 1882), pp. 43–45.

4. "Announcement," *Pop. Sci. Mo.*, 20 (1882), p. 409. "Literary Notice," *Pop. Sci. Mo.*, 20 (1882), pp. 557–60.

5. W. D. LeSueur, "Stallo's 'Concepts of Modern Physics,' " *Pop. Sci. Mo.*, 21 (1882), pp. 96–100, excerpted from the *Canadian Monthly*, 8 (1882), pp. 352–60. (This same volume of the *Canadian Monthly* also carried a brief review of Stallo's book on p. 324.) Stallo, "Speculative Science," pp. 145–64.

6. This latter anonymous review appeared in the Catholic journal *The Month*, 47 (1882), pp. 439–42.

7. Donald MacAlister, "Critical Notice," *Mind*, 8 (Apr. 1883), pp. 276–84. P. G. Tait, "Modern Physics," *Nature*, 8 (28 Sept. 1882), pp. 521–22. Newcomb, "Speculative Science," pp. 334–41. "Stallo's Physics," *The Nation*, 34 (1 June 1882), pp. 466–67; the anonymous author of this review was later identified to be G. Stanley Hall—see Haskell, *Indexes of Titles and Contributors to The*

[174]

Nation, I, p. 129. "Modern Physics," *The Critic,* 2 (25 Feb. 1882), p. 58. Judging from his apparent familiarity with physics, the anonymous author of this review in the *Critic* was likely Alfred M. Mayer, physicist and regular contributor to the *Critic.* The anonymous author, however, might also have been either chemist Ira Remsen or astronomer Charles A. Young, both also contributors. The editors listed these three scientists among the thirty or so persons who were "actual contributors" in 1882 to this journal of literature, science, the fine arts, music, and drama; this list appeared in the advertisement, "Now is the Time to Subscribe for *The Critic," The Critic,* 2 (30 Dec. 1882), p. 364.

8. Arnold W. Reinold, "Science," *The Academy,* 22 (18 Nov. 1882), pp. 366–67. Robert G. Eccles, "Review of Stallo's 'Concepts of Modern Physics,'" *Kansas City Review of Science and Industry,* 6 (Aug. 1882), pp. 229–39.

9. [Mayer], "Modern Physics," *The Critic,* 2 (25 Feb. 1882), p. 58. Mac-Alister, "Critical Notices," *Mind,* (8 Apr. 1883), p. 283. See also Tait, "Modern Physics," *Nature,* 8 (28 Sept. 1882) pp. 521–22.

10. Hall, "Stallo's Physics," *The Nation,* 34 (1 June 1882), p. 467.

11. Hall, "Stallo's Physics," *The Nation,* 34 (1 June 1882), p. 466. Le Sueur, "Stallo's 'Concepts,'" *Pop. Sci. Mo.,* 21 (1882), pp. 97–99. In a section of his original *Canadian Monthly* review (pp. 358–59) that was not reprinted in *Popular Science Monthly,* LeSueur cited other examples of the similarity between the ideas of Stallo and Comte.

12. Stallo, "Introduction to the Second Edition," pp. 40–41, n.21. Stallo, "The Primary Concepts of Modern Physical Science: Part IV—Inertia and Force," *Pop. Sci. Mo.,* 4 (1873–74), p. 352, n.1.

13. MacAlister, "Critical Notice," *Mind,* 8 (Apr. 1883), pp. 278–79. [Mayer], "Modern Physics," *The Critic,* 2 (25 Feb. 1882), p. 58.

14. Hall, "Stallo's Physics," *The Nation,* 34 (1 June 1882), pp. 466–67. Mac-Alister, "Critical Notice," *Mind,* 8 (Apr. 1883), p. 277.

15. Hall, "Stallo's Physics," *The Nation,* 34 (1 June 1882), p. 467. Reinold, "Science," *The Academy,* 22 (18 Nov. 1882), p. 367. See also Newcomb, "Speculative Science," pp. 337, 339; and Eccles, "Review of Stallo's 'Concepts,'" *Kansas City Review of Science and Industry,* 6 (Aug. 1882), p. 238.

16. MacAlister, "Critical Notice," *Mind,* 8 (Apr. 1883), pp. 279–80, 283–84.

17. Ibid., p. 284. Tait, "Modern Physics," *Nature,* 8 (28 Sept. 1882), p. 521.

18. [Mayer], "Modern Physics," *The Critic,* 2 (25 Feb. 1882), p. 58.

19. Bridgman, *The Logic of Modern Physics,* pp. 5, 6, 10.

20. Hall, "Stallo's Physics," *The Nation,* 34 (1 June 1882), pp. 466–67. Newcomb, "Speculative Science," pp. 334, 338–39.

3. THE LAWYER'S REBUTTAL

1. Stallo, *Concepts and Theories,* p. 10. Stallo, "Speculative Science," pp. 152, 155, 157.

2. Stallo, "Speculative Science," pp. 146, 148. Stallo, "Introduction to the Second Edition," pp. 27, 34–35.

3. Stallo, "Speculative Science," p. 153. Stallo, "Introduction to the Second Edition," p. 26.

4. Stallo, "Introduction to the Second Edition," pp. 23–24.

5. Stallo, "Speculative Science," pp. 161–62, 163. Stallo, "Introduction to the Second Edition," pp. 17–18, 47–48.

6. Stallo, *Concepts and Theories,* pp. 299, 306–07.

7. Stallo, "Introduction to the Second Edition," pp. 10, 16, 45–46.

8. Stallo, "Speculative Science," p. 157; *The Critic,* p. 58.

9. Stallo, "Introduction to the Second Edition," pp. 19–20, 28. For MacAlister's atomo-mechanical statements, see his "Critical Notice," *Mind,* 8 (Apr. 1883), pp. 281–282, 283, 279–80.

10. For recent favorable appraisals of Stallo's characterizations, see *Concepts and Theories,* ed. Bridgman, pp. xxi–xxii, and Drake, "J. B. Stallo and the Critique of Classical Physics," pp. 22–23.

4. THE ORTHODOXY OF MAYER AND DOLBEAR

1. This terminology is derived from that employed by Forman in "Weimar Culture, Causality, and Quantum Theory, 1918–1927," pp. 7, 38. See also the analytic distinctions made by Kuhn in discussing the "disciplinary matrix" and "paradigms" of a scientific community; Kuhn, "Postscript—1969," *Structure of Scientific Revolutions,* esp. pp. 181–91.

2. Alfred G. Mayer and Robert S. Woodward, "Alfred Marshall Mayer," *NAS Biog. Mem.,* 8 (1916), pp. 243–72.

3. Mayer, "Henry as a Discoverer," p. 475.

4. Mayer, *Lecture-Notes on Physics.* In this paragraph, I deal with pp. 3–4.

5. Mayer, *Lecture Notes on Physics,* pp. 89–91, 49–50.

6. Mayer, *The Earth a Great Magnet,* pp. 269–71.

7. Mayer, *Lecture-Notes on Physics,* pp. 5–11. Molella et al., eds., *A Scientist in American Life,* pp. 3–5, 36, 88–89.

8. Mayer and Barnard, *Light,* pp. 111–12. Mayer, *Earth a Great Magnet,* pp. 275–76.

9. Mayer, *Lecture-Notes on Physics,* pp. 5–11. See also Mayer, *Earth a Great Magnet,* p. 228.

10. Mayer, "On the Physical Conditions of a Closed Circuit," pp. 17, 19, 23. See also Mayer, *Earth A Great Magnet,* pp. 282–84.

11. Norton, "On Molecular Physics," pp. 61–73. Stallo, *Concepts and Theories of Modern Physics,* p. 114.

12. Thomson, *Electricity and Matter,* pp. 114–22. Prior to 1903, Thomson had been giving less formal expressions of his views regarding Mayer's model; see Snelders, "A. M. Mayer's Experiments," pp. 76–77. See also Strutt, *Life of Sir J. J. Thomson,* pp. 113–14, 139–40 and Heilbron, "J. J. Thomson and the Bohr Atom," pp. 23–24.

13. Alfred Mayer, "A Note on Experiments with Floating Magnets," *Amer. Jour. of Sci. and Arts,* 3d Ser. 15 (1878), pp. 276–77. See also "Scientific Intelligence," No. 10 "Floating Magnets" and No. 11 "Note on Floating Magnets," *Amer. Jour. of Sci. and Arts,* 3d Ser. 15 (1878), pp. 477–78. See also Snelders, "A. M. Mayer's Experiments," pp. 70–71.

14. Mayer, "On the Morphological Laws," pp. 254, 256.

15. [Alfred M. Mayer], "Modern Physics," *The Critic,* 2 (25 Feb. 1882), p. 58.

[176]

16. "Amos Emerson Dolbear," *National Cyclopaedia of American Biography,* IX, pp. 414–15.

17. Dolbear, *Matter, Ether, and Motion.* Dolbear added a large number of demonstrations illustrating "the vortex-ring theory of the constitution of matter" to his book, *The Art of Projecting.*

18. Dolbear, *Matter, Energy, and Motion,* pp. iii, 326, 49–50, 77–78.

19. Ibid., pp. vi, 311, 30, 43.

20. Ibid., pp. 279, 283, 297.

5. TROWBRIDGE AND ROWLAND: CAUTIOUS MECHANISTS

1. "List of Meetings of the Association," *Proceedings of the AAAS,* 52 (1902–03), p. 16; *Report of the Electrical Conference at Philadelphia in September, 1884,* pp. 7–12; Strutt, *Life of John William Strutt,* pp. 138–47; Thompson, *The Life of William Thomson,* II, pp. 810–39.

2. Hall, "John Trowbridge," in *NAS Biog. Mem.,* 14 (1932), pp. 185–204. Also see the unsigned, contemporary "Sketch of Professor John Trowbridge," *Pop. Sci. Mo.,* 26 (1884–85), pp. 836–39. Hall, "Physics: 1869–1928," pp. 277–91 and Hughes, "Engineering: 1847–1920," pp. 413–18 in *The Development of Harvard University,* ed. Morison. Hall, "John Trowbridge," pp. 526–27. Lyman, "Recollections [on the Jefferson Laboratory]," pp. 3–4; by permission of the Harvard University Archives.

3. Williams, "A History of Physics in Oberlin College," p. 53.

4. Trowbridge, "What is Electricity?" pp. 76–79, 87.

5. Ibid., pp. 77, 79, 83. Trowbridge, *The New Physics,* p. v.

6. Trowbridge earlier had studied vortex-rings; see "On Vortex-Rings in Liquids," pp. 131–36.

7. Trowbridge, "What is Electricity?" pp. 79, 77, 83–85. In 1878, he also sought to bring "the subject of the magnetic state of metals into the domain of mechanics"; see "On the Heat Produced by the Rapid Magnetization and Demagnetization of the Magnetic Metals," pp. 114–21.

8. Trowbridge, "What is Electricity?" pp. 81, 82, 85.

9. Letter from Trowbridge to Daniel Gilman, 12 Nov. 1882, reprinted in *Science in Nineteenth-Century America,* ed. Reingold, pp. 270–72. Strutt, *Life of John William Strutt,* pp. 145–57; see in particular the letter from Rayleigh to his mother, 19 Oct. 1884, p. 147. Hall, "Michelson and Rowland," p. 615.

10. Miller, "Rowland's Physics," pp. 39–45. Miller, "Rowland and the Nature of Electric Current," pp. 5–27 and "Rowland's Magnetic Analogy to Ohm's Law," pp. 230–41, both reprinted in *Science in America Since 1820,* ed. Reingold, pp. 186–220. Thomas C. Mendenhall, "Henry A. Rowland: Commemorative Address" (1901), reprinted in Rowland, *Physical Papers,* pp. 1–17.

11. Rowland, "Note on the Magnetic Effect of Electric Convection," *Phil. Mag.,* 7 (1879), reprinted in Rowland, *Physical Papers,* p. 138.

12. Letter from John Trowbridge to Daniel Gilman, 30 Nov. 1882, reprinted in *Science in Nineteenth-Century America,* ed. Reingold, pp. 272–74. See also Reingold, pp. 262–72.

13. Rowland, "A Plea for Pure Science," *Proc. AAAS,* 32 (1883), reprinted in Rowland, *Physical Papers,* pp. 596–97. Also reprinted during 1883 in *Science* and *Journal of the Franklin Institute.*

[177]

14. See Rowland, *Physical Papers,* pp. 635 (1884), 650 (1888), 659-60 (1889), 290 (1895), and 670 (1899).

15. Rowland, "On Modern Views with Respect to Electrical Currents," *Trans. Amer. Inst. Elec. Engineers,* 6 (1889), reprinted in Rowland, *Physical Papers,* pp. 569-60.

16. Rowland, ibid., p. 660.

17. Rowland, "On the Mechanical Equivalent of Heat . . . ," *Proc. Amer. Acad. Arts and Sci.,* 15 (1880), reprinted in Rowland, *Physical Papers,* p. 405. For Rowland's papers on heat, see *Physical Papers,* "Part III"; for his papers on light and spectra, see "Part IV."

18. Rowland, "The Electrical and Magnetic Discoveries of Faraday," *Elec. Rev.* (4 Feb. 1888), reprinted in Rowland, *Physical Papers,* pp. 638 and 651; "The Physical Laboratory in Modern Education," *Johns Hopkins Univ. Circulars,* no. 50 (1886), reprinted in *Physical Papers,* pp. 614-18. See also "On Atmospheric Electricity," *Johns Hopkins Univ. Circulars,* no. 19 (1882), reprinted in *Physical Papers,* p. 212.

19. Rowland, "Plea for Pure Science," reprinted in Rowland, *Physical Papers,* p. 608.

20. See, e.g., Rowland, "Research on the Absolute Unit of Electrical Resistance," *Amer. Jour. of Sci. and Arts,* 3d Ser., 15 (1878), reprinted in Rowland, *Physical Papers,* pp. 145-78.

21. Rowland, "Address as President of the Electrical Conference at Philadelphia, September 8, 1884," *Report of the Conference (1886),* reprinted in Rowland, *Physical Papers,* pp. 629-30. See also Rowland, "On Modern Views," reprinted in Rowland, *Physical Papers,* p. 660. Miller, "Rowland's Magnetic Analogy," p. 240 and "Rowland and the Nature of Electric Current," p. 20; Buchwald, "The Hall Effect and Maxwellian Electrodynamics in the 1880s; Part I: The Discovery of a New Electric Field," pp. 65-70. For Maxwell's shift in thinking, see Everitt, *James Clerk Maxwell,* pp. 101-02.

22. Rowland, "Electrical Conference," reprinted in Rowland, *Physical Papers,* pp. 630-31.

23. Rowland, "On Modern Views," reprinted in Rowland, *Physical Papers,* pp. 665-66. See also "Discoveries of Faraday," pp. 642-46.

24. Rowland, "Discoveries of Faraday," reprinted in Rowland, *Physical Papers,* p. 649.

25. Rowland, "Electrical Conference," reprinted in Rowland, *Physical Papers,* p. 631.

26. Ibid. See also, e.g., "On Modern Views," reprinted in Rowland, *Physical Papers,* p. 655.

27. Rowland, "Electrical Conference," reprinted in Rowland, *Physical Papers,* pp. 632-33. See also, e.g., "On Modern Views," reprinted in Rowland, *Physical Papers,* p. 655.

28. Rowland, "Electrical Conference," reprinted in Rowland, *Physical Papers,* p. 632.

29. Rowland, "On Modern Views," reprinted in Rowland, *Physical Papers,* pp. 658-59, 667.

30. Rowland, "Electrical Conference," reprinted in Rowland, *Physical Papers,* pp. 634-35.

6. MICHELSON AND HALL: EXPERIMENTATION AND EDUCATION

1. Livingston, *The Master of Light*; Jaffe, *Michelson and the Speed of Light*; Swenson, *The Ethereal Aether*; Shankland, "Michelson," pp. 19-25.
2. Michelson, "The Relative Motion of the Earth and the Luminiferous Ether," pp. 120-21. My discussion of this and Michelson's other ether writings reflects the historical insights of Swenson, Shankland, Jaffe, and Livingston as well as of Holton, "Einstein, Michelson, and the 'Crucial' Experiment," in *Thematic Origins of Scientific Thought,* pp. 261-352.
3. Michelson, "Relative Motion" (1881), pp. 128-29. Letter from Michelson to Bell, 17 April 1881, reprinted in *Science in Nineteenth-Century America,* ed. Reingold, pp. 288-90.
4. Letter from Michelson to Gibbs, 15 Dec. 1884, reprinted in *Science in Nineteenth-Century America,* ed. Reingold, pp. 307-08. Michelson and Morley, "Influence of Motion of the Medium on the Velocity of Light," pp. 377-86. Michelson also reported his results in a letter to Gibbs (reprinted in Swenson, *Ethereal Aether,* pp. 84-85) and in a letter to William Thomson (reprinted in Thompson, *Life of William Thomson,* II, p. 857).
5. Letter from Michelson to Rayleigh, 6 March 1887, reprinted in Strutt, *Life of John William Strutt,* pp. 343-45. Michelson and Morley, "On the Relative Motion of the Earth and the Luminiferous Ether," pp. 333-34, 341.
6. Michelson and Morley, "On the Relative Motion of the Earth and the Luminiferous Ether," p. 341. Michelson anticipated this entrapment rationalization earlier in the year when discussing the pending experiment with Rayleigh (Strutt, *Life of John William Strutt,* pp. 343-45) and in his 1884 letter to Gibbs (*Science in Nineteenth-Century America,* ed. Reingold, pp. 307-08).
7. Michelson, "A Plea for Light Waves," pp. 68, 76-78. (Michelson's address as vice-president of Section B, AAAS.) Joseph Lovering, "Michelson's Recent Researches on Light," *Smithsonian Inst. Annual Report,* (1889), pt. 2, p.468.
8. Bridgman, "Edwin Herbert Hall," *NAS Biog. Mem.,* 21 (1941), pp. 73-94. See also Edwin H. Hall Papers, Houghton Library, Harvard Univ.
9. Bridgman, "Edwin Hall," *NAS Biog. Mem.,* 21 (1941), pp. 84-85. Miller, "Rowland and the Nature of Electric Currents," pp. 8, 27, and "Rowland's Magnetic Analogy to Ohm's Law," p. 241. Buchwald, "The Hall Effect and Maxwellian Electrodynamics," pp. 70-91.
10. Sopka, "The Discovery of the Hall Effect," in *The Hall Effect and Its Applications,* ed. Chien and Westgate, pp. 523-45. Buchward, "The Hall Effect and Maxwellian Electrodynamics," pp. 70-91. Bridgman, "Edwin Hall," *NAS Biog. Mem.,* 21 (1941), pp. 79-80. Miller, "Rowland and the Nature of Electric Currents," pp. 18-21. Hall disagreed with Maxwell's contention in *Electricity and Magnetism* that a magnetic force acts only on the *conductor* that carries the electric current as distinguished from the *current* itself.
11. Hall, "On the New Action of Magnetism on a Permanent Electric Current," pp. 184-85.
12. Quoted in Bridgman, "Edwin Hall," *NAS Biog. Mem.,* 21 (1941), p. 82. See also Sopka, "Discovery of the Hall Effect," p. 523. Thompson, *Life of William Thomson,* pp. 807-08, 1257; Hall, "On the Rotation of the Equipotential Lines of an Electric Current by Magnetic Action," p. 131.
13. Hall, "Experiments on the Effect of Magnetic Force on the Equipotential Lines of an Electric Current," p. 139.

14. Moyer, "Edwin Hall and the Emergence of the Laboratory in Teaching Physics," pp. 96–103. See also Rosen, "A History of the Physics Laboratory in the American Public High School," pp. 194–204.

15. Hall, "Teaching Elementary Physics," p. 331.

7. MESSAGES FROM EUROPE

1. Kevles, "The Physics, Mathematics, and Chemistry Communities," in *The Organization of Knowledge in Modern America, 1860–1920,* eds., Oleson and Voss, p. 139 and tables 1, 10. Kevles, "On the Flaws of American Physics," in *Nineteenth-Century American Science,* ed. Daniels, p. 140.

2. "Speech of Prof. Tyndall," in *Proceedings at the Farewell Banquet to Professor Tyndall,* p. 49.

3. "Professor von Helmholtz's Visit to America," *The Physical Review,* 1 (1894), p. 229; Edward L. Nichols, "Hermann von Helmholtz," *Phys. Rev.,* 2 (1895), p. 227; Thomas C. Mendenhall, "Helmholtz," *Annual Report of Smithsonian Inst.,* (1895), p. 793.

4. Turner, "Hermann von Helmholtz," *Dictionary of Scientific Biography,* pp. 250–52; hereafter cited as *DSB.* See also Helmholtz, *Epistemological Writings.*

5. Kevles, "Physics, Mathematics, and Chemistry Communities" in *The Organization of Knowledge in Modern America,* eds. Oleson and Voss, tables 5, 6.

6. Mayer, *Earth a Great Magnet,* pp. 260–61, 276; Trowbridge, "What is Electricity?" p. 83; and Rowland, "The Electrical and Magnetic Discoveries of Faraday," reprinted in Rowland, *Physical Papers,* pp. 649–50.

7. Williams, *Michael Faraday,* pp. 506–13.

8. Hall, "Inertia," *Science,* 3 (1884), pp. 482–83; for criticism of this note and Hall's rebuttal, see "Letters to the Editor," *Science,* 3 (1884), pp. 559–62.

9. Tyndall, *Lectures on Light,* pp. 3–11; *Proceedings at the Farewell Banquet to Professor Tyndall,* pp. 3–9. Sopka, "John Tyndall," pp. 193–203. Sopka, "An Apostle of Science Visits America," pp. 369–75.

10. Eve and Creasey, *Life and Work of John Tyndall,* p. 167; Tyndall, *Lectures on Light,* pp. 5–6, 192.

11. Stallo, *Concepts and Theories,* p. 154. Tyndall, *Lectures on Light,* p. 34.

12. Tyndall, *Lectures on Light,* pp. 69–70, 104, 164.

13. For Thomson's American travels and for Gilman's letter to Thomson, Nov. 1882, see Thompson, *Life of William Thomson,* pp. 806–39. See also Sharlin, *Lord Kelvin,* pp. 186–87, 202–12.

14. William Thomson, "The Wave Theory of Light," pp. 263–86. William Thomson, *Notes of Lectures on Molecular Dynamics and the Wave Theory of Light*; "papyrograph" reproduction, 1884, courtesy of the Rare Book Dept., University of Wisconsin Library. Two decades later, Thomson (by then Lord Kelvin) published a revised version of these *Notes of Lectures*; see *Baltimore Lectures on Molecular Dynamics and the Wave Theory of Light.* For a technical commentary on this later version of the *Lectures,* see Doran, "Origins and Consolidation of Field Theory," pp. 133–260.

15. William Thomson, *Notes of Lectures* (1884), pp. 1, 2, 5. For a summary of the lectures, see George Forbes's three notes in *Nature,* 31 (1885), pp. 461–63,

[180]

508-10, 601-03; see also Thompson, *Life of William Thomson,* pp. 816-36, 1035-38 and Sharlin, *Lord Kelvin,* chapter 12.

16. Henry Crew Diary entry of 14 Oct. 1884, courtesy of Spencer R. Weart, American Institute of Physics. William Thomson, "Wave Theory of Light," pp. 268, 277.

17. William Thomson, *Notes of Lectures,* pp. 6-10.

18. William Thomson, "Wave Theory of Light," pp. 286, 277. Cf. Swenson, *The Ethereal Aether,* pp. 77-80.

19. William Thomson, *Notes of Lectures,* pp. 270-71, 131-32, 10. In the 1904 edition of the *Baltimore Lectures,* Thomson deleted these passages in the eleventh and twentieth chapters that extolled mechanical models; see pp. 131, 436. Bridgman, in *Logic of Modern Physics,* p. 45, misinterprets Thomson's statements.

20. For a discussion of Maxwell's shift in thinking from molecular vortices or forces between electric particles toward general dynamical principles founded on electrical experiments, see Everitt, *James Clerk Maxwell,* esp. pp. 101-02.

21. William Thomson, *Notes of Lectures,* pp. 132, 271, 5-6.

22. Thompson in his *Life of William Thomson,* p. 819, n. 1, wrote: "This passage and others of similar import led his audience to conclude that at that date the lecturer had never read Clerk Maxwell's book!" For Thomson's views on Maxwell's theory, see also Buchwald, "Sir William Thomson," *DSB* (1976), XIII, pp. 384-85.

23. Duhem, *The Aim and Structure of Physical Theory,* pp. 70, 74, 81, 84-5.

24. Letter from Lady Thomson to George Darwin, 7 Oct. 1884, reprinted in Thompson, *Life of William Thomson,* p. 809; letter from Lord Rayleigh to Lady Rayleigh, 19 Oct. 1884, reprinted in Strutt, *Life of John William Strutt,* pp. 146-47. Henry Crew Diary entry of 17 Oct. 1884. Livingston, *Master of Light,* p. 107; Livingston also cites Crew's diary.

8. NEWCOMB'S OPERATIONAL OUTLOOK

1. W. W. Campbell, "Simon Newcomb," *Memoirs of the NAS,* 17, 1st. Mem. (1924), pp. 1-8, bibliography, pp. 19-69; Newcomb, *Reminiscences of an Astronomer*; and Norberg, "Simon Newcomb's Early Astronomical Career," pp. 209-25.

2. Diary, 5 March 1860 through 20 Sept. 1860 and 2 Feb. 1861 through 18 March 1861, Simon Newcomb Papers, Manuscript Division, Library of Congress, Box 1. See also Norberg, "Simon Newcomb's Early Astronomical Career," pp. 209-25.

3. For a detailed discussion of the development of Newcomb's operational perspective, see Moyer, "In Advocacy of Science: The Essays of Simon Newcomb," forthcoming.

4. Madden, *Chauncey Wright and the Foundations of Pragmatism,* pp. 3-30. The 1878 quotation about Wright is from a letter of Newcomb to James B. Thayer, in *Letters of Chauncey Wright,* p. 70. See also Carolyn Eisele, "The Charles S. Peirce-Simon Newcomb Correspondence," pp. 409-33.

5. Newcomb, "Exact Science in America," p. 293.

6. Wright, "Speculative Dynamics," *The Nation* (June 1875), reprinted in

Philosophical Discussions, ed. Norton, pp. 385, 388–89. Wright, "The Philosophy of Herbert Spencer," *North American Review* (Jan. 1865), reprinted in *Philosophical Discussions,* pp. 43–96, esp. pp. 46–47, 78. See also Madden, *Chauncey Wright,* pp. 73–81 and Wiener, *Evolution and the Founders of Pragmatism,* pp. 70–96.

7. Peirce, "How To Make Our Ideas Clear," *Pop. Sci. Mo.* (Jan. 1878), reprinted in *Philosophical Writings of Peirce,* ed. Buchler, pp. 31, 35–36. See also *Charles S. Peirce, Selected Writings,* ed. Wiener, p. 113.

8. Newcomb, *Reminiscences of an Astronomer,* pp. 70–71. Newcomb, "Abstract Science in America, 1776–1876," *North American Review,* 122 (1876), p. 110.

9. Along with the *Popular Science Monthly Supplement* and the London-based *Journal of Science,* the journals included the *American Association Proceedings,* the *Independent,* and the *Kansas City Review*; see Campbell's bibliography (n. 1) for full documentation.

10. Newcomb, "The Course of Nature," pp. 481–82.

11. Besides the pamphlet published by Judd and Detweiler and the book published by Harper and Bros., the speech was reprinted in the *Bull. Phil. Soc. Wash.* and the *Smithsonian Misc. Coll.*; see Campbell's bibliog. (n. 1) for full documentation.

12. Newcomb, "The Relation of Scientific Method to Social Progress," pp. 319, 224–26.

13. Cf. Newcomb, "Relation of Scientific Method to Social Progress," pp. 327–28, with Wright, "Evolution of Self-Consciousness," *North Amer. Rev.* (April 1873), reprinted in *Philosophical Discussions,* ed. Norton, pp. 244–46.

14. Cf. Newcomb, "Relation of Scientific Method to Social Progress," p. 320, with Wright, "Speculative Dynamics," reprinted in *Philosophical Discussions,* ed. Norton, p. 387.

15. Newcomb, "Speculative Science," p. 339.

16. See, e.g., pp. 386 and 393 of Wright's "Speculative Dynamics," reprinted in *Philosophical Discussions,* ed. Norton, p. 387.

17. Newcomb, "Law and Design in Nature," *North American Review,* 28 (May 1879), p. 542. See also chapters 1–4 of Wiener, *Evolution and the Founders of Pragmatism.*

18. Newcomb, "Modern Scientific Materialism," p. 1, cols. 1–4. Newcomb, "Course of Nature," p. 484.

19. Peirce, "The Order of Nature," *Pop. Sci. Mo.,* (June 1878), reprinted in *Charles S. Peirce: Essays,* ed. Tomas, pp. 107, 122–23; see also "How To Make Our Ideas Clear," in *Philosophical Writings of Peirce,* ed. Buchler, pp. 33, 35, 38–39. William L. Rosensohn has traced Peirce's devotion to and eventual break with (in the mid-1880s) the mechanical philosophy; see Rosensohn, *The Phenomenology of Charles S. Peirce,* pp. 62–75. The final quotation is from Peirce, "Concerning the Author," ms. (c. 1897), reprinted in *Philosophical Writings,* ed. Buchler, p. 1.

20. Mach instigated—and then wrote a generally complimentary introduction to—the 1901 German publication of Stallo's book. Mach's relation to Stallo is discussed by Bridgman in his introduction to the 1960 edition of Stallo's book, pp. viii and xvi. In his introduction, Bridgman also endorsed Stallo's endeavor, p. xxi.

21. Stallo, *Concepts and Theories of Modern Physics,* pp. 8, 201.

22. Newcomb, "Speculative Science," pp. 335–36, 338.

23. Cf. Newcomb, "Speculative Science," p. 339, with Peirce, "How To Make Our Ideas Clear," pp. 31-32.
24. Stallo, "Speculative Science," pp. 147-48.
25. Newcomb, "Speculative Science," p. 340.
26. Stallo, "Speculative Science," pp. 153, 154, 157.

9. GIBBS'S PHENOMENALISTIC LEANINGS

1. Wheeler, *Josiah Willard Gibbs*; Rukeyser, *Willard Gibbs*; Martin J. Klein, "Josiah Willard Gibbs," *DSB*; Seeger, *J. Willard Gibbs*; Charles Hastings, "Josiah Willard Gibbs," *NAS. Biog. Mem.*, 6 (1909), pp. 373-93; Henry Bumstead, "Josiah Willard Gibbs," *Amer. Jour. Sci.* (1903), reprinted and enlarged in Gibbs, *The Collected Works*, I, pp. xxiii-xxviii. My subsequent discussion of Gibbs's writings reflects the insights of these biographers.
2. Letter from Rowland to Gibbs, 3 March 1879, reprinted in Wheeler, *Gibbs*, p. 87; letter from Rowland to Gilman, 8 May 1879, reprinted in *Science in Nineteenth-Century America*, ed. Reingold, pp. 317-18.
3. Letter from Michelson to Gibbs, 15 Dec. 1884, reprinted in Wheeler, *Gibbs*, pp. 140-41; letter from Rowland to Gibbs, 29 July 1884, quoted in Wheeler, *Gibbs*, p. 142. For a description of letters concerning Trowbridge and Stallo, see Wheeler, App. III, "Catalogue of the 'Scientific Correspondence,'" *Gibbs*, pp. 221-22.
4. The following articles are reprinted in *The Collected Works of J. Willard Gibbs*, Vol. I: "Graphical Methods in the Thermodynamics of Fluids," *Trans. Conn. Acad.*, 2 (1873), pp. 1-32; "A Method of Geometrical Representation of the Thermodynamic Properties of Substances by Means of Surfaces," *Trans. Conn. Acad.*, 2 (1873), pp. 33-54; "On the Equilibrium of Heterogeneous Substances," *Trans. Conn. Acad.*, 3 (1875-76 and 1877-78), pp. 53-353; and "Abstract of the Preceding Paper," *Amer. Jour. Sci.*, 3d Ser. 16 (1878), pp. 354-71. For Gibbs's Rumford Medal comment, see letter from Gibbs to American Academy of Arts and Sciences, 10 Jan. 1881, reprinted in *A Commentary on the Scientific Writings of J. Willard Gibbs*, eds. Donnan and Haas, pp. 54-55.
5. Klein, "Gibbs on Clausius," pp. 129-35.
6. Gibbs, "Graphical Methods," p. 2,n.; "Geometrical Representation," p. 52,n.; "Equilibrium of Heterogeneous Substances," p. 55; and "Abstract," p. 354. See full citation in n.4; these articles are reprinted in *The Collected Works of J. Willard Gibbs*, Vol. I.
7. Klein, "Gibbs," *DSB*, p. 391; Wheeler, *Gibbs*, pp. 99-102. See also Klein, "Mechanical Explanation at the End of the Nineteenth Century," pp. 76-78.
8. Gibbs, "Equilibrium of Heterogeneous Substances," pp. 166-67, reprinted in *The Collected Works of J. Willard Gibbs*, Vol. I. For a discussion of statistical interpretations of the second law, see Klein, "Gibbs on Clausius," pp. 144-45.
9. Gibbs, "Rudolf Julius Emanuel Clausius," *Proc. Amer. Acad.*, NS 16 (1889), reprinted in *Collected Works*, II, pp. 262-65. See also Klein, "Gibbs on Clausius," pp. 127-49.
10. Gibbs, "Abstract: On the Fundamental Formula of Statistical Mechanics . . .," *Proc. AAAS*, 33 (1884), reprinted in *Collected Works*, II, 16. See also Wheeler, *Gibbs*, p. 154. Letter from Gibbs to Rayleigh, 27 June 1892, quoted in Klein, "Gibbs," *DSB*, pp. 391-92.

[183]

11. For a discussion of these papers, see Wheeler, *Gibbs,* chapter VIII, pp. 121–33. See also Leigh Page, "Gibbs' Contributions to the Theory of Light," in *Commentary on Scientific Writings,* II, ed. Donnan and Haas, pp. 113–26.

12. Gibbs, "On the General Equations of Monochromatic Light in Media of Every Degree of Transparency," *Amer. Jour. Sci.,* 3d ser. 25 (1883), reprinted in *Collected Works,* II, p. 212,n. Gibbs, "On Double Refraction and the Dispersion of Colors in Perfectly Transparent Media," *Amer. Jour. Sci.,* 3d ser. 23 (1882), reprinted in *Collected Works,* II, p. 182.

13. Gibbs, "On Double Refraction in Perfectly Transparent Media Which Exhibit the Phenomena of Circular Polarization," *Amer. Jour. Sci.,* 3d ser. 23 (1882), reprinted in *Collected Works,* II, p. 210.

14. Gibbs, "A Comparison of the Elastic and the Electrical Theories of Light with Respect to the Law of Double Refraction and the Dispersion of Colors," *Amer. Jour. Sci.,* 3d ser. 35 (1888), reprinted in *Collected Works,* II, pp. 223–24, 231.

15. Gibbs, "A Comparison of the Electric Theory of Light and Sir William Thomson's Theory of a Quasi-Labile Ether," *Amer. Jour. Sci.,* 3d ser. 37 (1889), reprinted in *Collected Works,* II, p. 232, 245–46.

10. LANGLEY AND NIPHER: SKEPTICS

1. White, Pickering, and Chanute, "Samuel Pierpont Langley," pp. 1–49; Vaeth, *Langley*; and Obendorf, "Samuel P. Langley."

2. Adams, *The Education of Henry Adams,* p. 377. Letter from Langley to Peirce, 3 April 1901, reprinted in Wiener, "The Peirce-Langley Correspondence," p. 205.

3. For Nipher's biography, see: "Francis Eugene Nipher," *Dict. Amer. Biog.* (1934), XIII, pp. 525–26, and *Nat. Cyclopaedia Amer. Biog.* (1932), XXII, pp. 55–56.

4. Langley, *The New Astronomy,* pp. 175–76, 249–51.

5. "Notes and News," *Science,* 10 (1887), pp. 80–81. Langley, "The History of a Doctrine," pp. 1–23.

6. Langley, "History of a Doctrine," pp. 2, 12, 13, 21, 23.

7. Nipher, *Physical Law,* pp. 1–7.

8. Ibid., pp. 8–9.

9. Langley, "History of a Doctrine," pp. 7, 18–19; Langley, *New Astronomy,* pp. 70–72. In earlier years, Langley more implicitly accepted atomo-mechanical presuppositions; see, e.g., "The Recent Progress of Solar Physics," *Pop. Sci. Mo.,* 16 (1879), pp. 7–8.

10. Nipher, "The Ether," pp. 127–30, 133. Nipher, *Electricity and Magnetism.*

11. Stallo, *Concepts and Theories of Modern Physics,* p. 295; "Introduction to the Second Edition," p. 16; "Primary Concepts of Modern Physical Science," p. 108. Bridgman, in his introduction to Stallo's *Concepts and Theories,* 3d ed., p. xxvii, comments on Stallo's "Greek" attitude that "the external world is friendly to human reason."

11. A REAPPRAISAL OF STALLO AND HIS CRITICS

1. Stallo, *Concepts and Theories of Modern Physics*, p. 23.
2. For similar conclusions regarding international physics, see Harman, *Energy, Force and Matter*, pp. 1-11; Brush, *The Kind of Motion We Call Heat*, I, pp. 35-103; and Swenson, *The Ethereal Aether*, pp. 77, 89.
3. Stallo, *Concepts and Theories*, p. 10.
4. Stallo, "Speculative Science," p. 152.
5. Ibid, pp. 161-63.

12. REALIGNMENTS WITHIN THE OLD GUARD

1. Joseph Ames, review of *Baltimore Lectures on Molecular Dynamics and the Wave Theory of Light*, by Lord Kelvin, *Phys. Rev.*, 19 (1904), pp. 62-64.
2. During the spring of 1902, Kelvin returned to the United States; his visits to Columbia, Cornell, Yale and other universities were largely ceremonial. See Thompson, *Life of William Thomson*, II, pp. 1165-69; Sharlin, *Lord Kelvin*, pp. 233-34.
3. For overviews of physics circa 1900, see Hiebert, "The State of Physics at the Turn of the Century," pp. 3-22; Klein, "Mechanical Explanation at the End of the Nineteenth Century," pp. 58-82; Brush, *The Kind of Motion We Call Heat*, I, pp. 35-103; and McCormmach, "H. A. Lorentz and the Electromagnetic View of Nature," esp. pp. 485-97.
4. Edward Nichols, "Henry Augustus Rowland, Physicist," *Phys. Rev.*, 13 (1901), p. 62.
5. Rowland, "The Röntgen Ray, And Its Relation to Physics: A Topical Discussion," *Trans. Amer. Inst. of Elec. Eng.*, 13 (1896), reprinted in *Physical Papers*, p. 581. Cf. the first paragraph of this quotation to Stallo, *Concepts and Theories of Modern Physics*, pp. 114-15.
6. Rowland, "The Highest Aim of the Physicist," *Science*, 10 (1899), reprinted in *Physical Papers*, pp. 674-76. See also Miller, "Rowland's Physics," pp. 44-45.
7. See, e.g., Ernest Meritt, "On Kathode Rays and Some Related Phenomena, II," *Science*, NS 12 (1900), p. 102; and Edward Nichols, "The Fundamental Concepts of Physical Science," in *Inorganic Sciences*, Vol. IV of *Congress of Arts and Science*, ed. Rogers, p. 21.
8. Trowbridge, *What Is Electricity?* p. 3.
9. Cf. Trowbridge, *What Is Electricity?* (1896), esp. pp. 1-10, 264-86, 305-09, to Trowbridge, "What Is Electricity?" *Pop. Sci. Mo.*, 26 (1884), pp. 76-88.
10. Cf. Trowbridge, *What Is Electricity?*, p. 82, to Stallo, *Concepts and Theories*, p. 307.
11. Frank Whitman, rev. of *What Is Electricity?* by Trowbridge, *Phys. Rev.*, 5 (1897), pp. 61-63. Trowbridge and Rollins, "Radium and the Electron Theory," pp. 77-79. For Trowbridge's later and even fuller adoption of new trends, see his "Physical Science of Today," pp. 318-24.
12. Langley, "The Laws of Nature," pp. 922, 924.
13. Ibid., pp. 922, 926-27.
14. Ibid., p. 923. Letter from Langley to Peirce, 19 Apr. 1901, reprinted in Wiener, "The Peirce-Langley Correspondence," pp. 206-07.

15. Wiener, "Peirce-Langley Correspondence," pp. 201–14.

16. The following page numbers refer to Peirce's reprinted articles in *Charles S. Peirce*, ed. Wiener: "Science and Immortality," *Christian Register* (Apr. 1887), pp. 348–50; "The Architecture of Theories," *The Monist* (Jan. 1891), pp. 146–48; and "The Doctrine of Necessity," *The Monist* (Apr. 1892), p. 171. See also *Charles S. Peirce*, ed. Wiener, p. 142; and Forman, "Weimar Culture, Causality, and Quantum Theory, 1918–1927," pp. 66–67.

17. See Peirce "What Pragmatism Is," *The Monist*, 15 (1905), reprinted in *Selected Writings*, ed. Wiener, pp. 180–202. James, "What Pragmatism Means," from *Pragmatism* (1907) reprinted in *The American Pragmatists*, eds. Konvitz and Kennedy, p. 30.

18. *The Nation*, 68 (Feb. 1899), pp. 95–96. Peirce was identified to be the anonymous author of this review by Haskell, *Indexes of Titles and Contributors to The Nation*. For another negative review of Holman's book, see J. G. MacGregor, *Phys. Rev.*, 9 (1899), pp. 59–64.

19. Holman, *Matter, Energy, Force and Work*, pp. 225–26. The letter from Kelvin to Holman (18 May 1898) as well as the demise of the vortex-atom theory are discussed in Silliman, "William Thomson," pp. 461–74.

20. Reprinted in *Collected Works of J. Willard Gibbs*, II, esp. pp. vii–xii, 165–67.

21. For an explication of Gibbs's statistical mechanics, see Haas, "Gibbs and the Statistical Conception of Physics," in *Commentary on the Scientific Writings of J. Willard Gibbs*, ed. Donnan and Haas, II, pp. 161–78. See also Klein, *Paul Ehrenfest*, pp. 128–37; and Brush, *The Kind of Motion We Call Heat*, I, pp. 94, 98–99.

13. RETRENCHMENT

1. Michelson, "The Relative Motion of the Earth and the Ether," pp. 475–78; Michelson, "A Theory of the 'X-Rays,'" pp. 312–14.

2. Michelson, *Light Waves*, pp. 2, 23–24. Badash, "The Completeness of Nineteenth-Century Science," p. 52.

3. Michelson, *Light Waves*, pp. 46, 112–18, 148, 161–62. Loyd Swenson feels that Michelson's interjection in the quotation—"it ought to be true even if it is not"—implies a wavering in his mechanistic faith; see Swenson, *Ethereal Aether*, p. 128.

4. Dolbear, *Modes of Motion*, pp. 92, 104–06, 29–32. Dolbear, *First Principles of Natural Philosophy*, esp. pp. iii–v. K. E. Guthe, review of *First Principles*, *Phys. Rev.*, 5 (1897), pp. 186–87. Dolbear also had a third 1897 publication, an expanded lecture titled, *The Machinery of the Universe: Mechanical Conceptions of Physical Phenomena*.

5. Hall, "A Tentative Theory of Thermo-Electric Action," pp. 313–88.

6. Hall, "Physics Teaching at Harvard Fifty Years Ago," pp. 17, 19. For a similar comment on force and energy, see Hall, "John Trowbridge," *NAS Biog. Mem.*, 14 (1932), p. 190. See also Hall, "Physics: 1869–1928," in *Development of Harvard University*, ed, Morison, p. 278.

7. Mayer and Woodward, "Alfred Marshall Mayer," *NAS Biog. Mem.*, 8 (1916), pp. 260–62.

[186]

8. Nipher, "Present Problems in the Physics of Matter," in *Congress of Arts and Science,* IV, pp. 87–88.

9. Ibid., pp. 98–101, 92. See also Nipher, "Physics During the Last Century," pp. 105–23.

10. Newcomb, "The Coming International Congress of Arts and Science at St. Louis, September 19–24," pp. 466–68.

11. Newcomb, *His Wisdom, the Defender,* pp. 12, 87, 122, 150.

14. THE ST. LOUIS CONGRESS: FOREIGN DELEGATES

1. Lovejoy, review of *Congress of Arts and Science, Universal Exposition, St. Louis, 1904,* ed. Rogers, *Science,* 23 (1906), p. 655. Review of *Congress of Arts and Science,* ed. Rogers, *The Nation,* 82 (June 1906), p. 476. Peirce was later identified to be the author of this anonymous review; see Haskell, *Indexes of Titles and Contributors to The Nation.* Davis, "The International Congress," p. 8.

2. "Purpose and Plan of the Congress," in *Philosophy and Mathematics,* Vol. I of *Congress of Arts and Science,* ed. Rogers, pp. 50–51. Hereafter cited as *Congress* I. See also the other articles at the beginning of this volume, pp. 1–84, on "The History of the Congress." Also see Simon Newcomb, "The International Congress of Scholars at St. Louis," *The Nation,* 79 (Sept. 1904), p. 193.

3. Münsterberg, "The Scientific Plan of the Congress," in *Congress* I, esp. pp. 125–26. "Proceedings of the APS: Minutes of the 25th Meeting," *Phys. Rev.,* 19 (1904), p. 298.

4. Poincaré, "The Principles of Mathematical Physics," trans. George Halsted, in *Congress* I, pp. 606–08, 617.

5. Ibid., pp. 609–10, 615, 621.

6. Ibid., pp. 610–12.

7. See Bernstein, *Einstein,* pp. 87–87; Holton, "Poincaré and Relativity," in *Thematic Origins of Scientific Thought,* pp. 185–95; and Swenson, *Ethereal Aether,* pp. 149–51. Lovejoy, review of *Congress of Arts and Science, St. Louis, 1904,* ed. Rogers, *Science,* 23 (1906), p. 659.

8. Poincaré, "Principles of Mathematical Physics," *Congress* I, pp. 605–06, 621, 604.

9. Langevin, "The Relations of Physics of Electrons to Other Branches of Science," trans. Bergen Davis, in *Inorganic Sciences,* Vol. IV of *Congress of Arts and Science,* pp. 122–28, 156, 138. Hereafter cited as *Congress* IV.

10. Ibid., p. 156. See also Hiebert, "State of Physics at the Turn of the Century," pp. 15–17.

11. Letter from Rutherford to Boltwood, 9 Sept. 1904, reprinted in *Rutherford and Boltwood: Letters on Radioactivity,* ed. Badash, p. 43. Rutherford, "Present Problems of Radioactivity," in *Congress* IV, pp. 159 and 171. See also Hiebert, "State of Physics at the Turn of the Century," p. 18.

12. Arthur G. Webster, "Some Practical Aspects of the Relations Between Physics and Mathematics," *Phys. Rev.,* 18 (1904), p. 304. For examples of papers read at American meetings by Rutherford, see *Phys. Rev.,* 13 (1901), pp. 321–44; and 16 (1903), pp. 181–83. For his election to the APS Council, see "Proceedings of the APS," *Phys. Rev.,* 18 (1904), pp. 116–17. See also Badash, "The Origins

of Big Science," in *Rutherford and Physics at the Turn of the Century,* ed. Bunge and Shea, pp. 29-30; and *Radioactivity in America,* pp. 28-29, 54-55.

13. Millikan, "Abstract of 'The Relation Between the Radioactivity and the Uranium Content of Certain Minerals,'" in *Congress* IV, p. 187. For Millikan's comments on the Congress, see his *Autobiography of Robert A. Millikan,* pp. 22, 84-85. For his overview of recent happenings in physics, see the chapter on "Invisible Radiations" in Millikan and Gale, *A First Course in Physics,* pp. 461-82. See also Badash, *Rutherford and Boltwood,* p. 12; "The Origins of Big Science," pp. 29-30; and *Radioactivity in America,* pp. 63-64, 78-80.

14. James, *Pragmatism,* partially reprinted in *The American Pragmatists,* eds. Konvitz and Kennedy, pp. 30, 34.

15. Ostwald, "On the Theory of Science," trans. R. M. Yerkes, in *Congress* I, pp. 347-49.

16. *Phys. Rev.,* 1 (1893-94), pp. 233-35. See also review of *Manual of Physico-Chemical Measurements,* by Ostwald, *Phys. Rev.,* 2 (1894-95), pp. 395-97.

17. Buckingham, review of *Grundzüge der Mathematischen Chemie, Energetik der Chemischen,* by Helm, *Phys. Rev.,* 3 (1895-96), pp. 152-54; see also Buckingham's plea for energetics in his review of a book by Nernst, in *Phys. Rev.,* 4 (1896-97), p. 156. Millikan, *Autobiography,* pp. 21-22.

18. Davis, "International Congress," p. 22. Millikan, *Autobiography,* pp. 22, 84-85.

19. Ludwig Boltzmann, "The Relations of Applied Mathematics," trans. S. Epsteen, in *Congress* I, pp. 592-95, 599, 601-03. Boltzmann lectured the following summer at Berkeley; see "Reise eines deutschen Professors ins Eldorado."

20. These sets of lectures were published in the United States as: *Electricity and Matter* and *The Discharge of Electricity through Gases.* See also Thomson, *Recollections and Reflections.*

21. Lorentz, *The Theory of Electrons,* pp. 79-80, 223-30. See also Hass-Lorentz, ed., *H. A. Lorentz.*

22. Brush, "Mach," in *The Kind of Motion We Call Heat,* pp. 274-99. Cohen and Seeger, eds., *Ernst Mach.* See also Mott, *History of American Magazines,* pp. 302-03. McCormack, "John Bernard Stallo," pp. 276, 283.

15. AMERICANS AT ST. LOUIS

1. Biographical sketches of Barus, Kimball, Brace, and Nichols appear with their St. Louis lectures in *Congress* IV. Also see R. B. Lindsay, "Carl Barus," *NAS Biog. Mem.,* 22 (1941), pp. 171-213; C. A. S., "Dewitt Brace," *Dict. Amer. Biog.,* (1929); and Ernest Merritt, "Edward Nichols," *NAS Biog. Mem.,* 21 (1940), pp. 343-66. In addition, see Knowlton, "Henry Crew," pp. 168-74; Lee Dubridge and Paul Epstein, "Robert Millikan," *NAS Biog. Mem.,* 33 (1959), pp. 241-82; and Millikan, *Autobiography.*

2. See the 1904 exchange of letters between Michelson and George Hale, quoted in Livingston, *Master of Light,* pp. 228-29.

3. Woodward, "The Unity of Physical Science," *Congress* IV, p. 5. Erwin Hiebert sees Woodward's comments as evidence of a resurgence of atomic-molecular ideas following the discovery of X rays and other spectacular phenomena; see Hiebert, "State of Physics at the Turn of the Century," p. 8.

4. Barus, "The Progress of Physics in the Nineteenth Century," *Congress* IV, pp. 64–65, 38.

5. Kimball, "The Relations of the Science of Physics of Matter to Other Branches of Learning," *Congress* IV, pp. 74–76.

6. Brace, "The Ether and Moving Matter," *Congress* IV, pp. 113–17.

7. Nichols, "The Fundamental Concepts of Physical Science," *Congress* IV, pp. 24–27.

8. Kimball, "Relations of the Science of Physics," *Congress* IV, pp. 74–76. Kelvin's 1900 lecture on "Nineteenth Century Clouds Over the Dynamical Theory of Heat" is reprinted in his *Baltimore Lectures,* pp. 486–527. Brace, "Ether and Moving Matter," *Congress* IV, pp. 116–17.

9. Barus, "Progress of Physics," *Congress* IV, pp. 62–64. Kimball, "Relations of the Science of Physics," *Congress* IV, pp. 72–73.

10. Crew, "Recent Advances in the Teaching of Physics," pp. 486–87.

11. Ibid., pp. 484–86. Crew, "What Can Be Done to Make the Study of Physics a Better 'Training for Power?'" pp. 523–26.

12. Bridgman, *Dimensional Analysis,* pp. vii, 26–27, 51. See also Everitt, *Maxwell,* p. 100.

13. Nichols, "Fundamental Concepts of Physical Science," *Congress* IV, pp. 18–21.

14. Ibid., pp. 18–21, 27.

16. FRANKLIN'S OPERATIONAL PERSPECTIVE

1. "William Suddards Franklin," *Nat. Cyc. Amer. Biog.* (1932), XXII, p. 140. Wilson, "William Suddards Franklin," pp. 162–64. A few of Franklin's private papers are in the Institute Archives and Special Collections at MIT, while a few biographical items are in the MIT Historical Collections. See also Phillips, "History of the Association," pp. 49, 51.

2. Franklin, "Popular Science," pp. 361–62. This speech was reprinted in a special lead section, for the AAAS meeting, of *Science,* 17 (1903), pp. 8–15.

3. Edwin Hall to William James, 13 Oct. 1908 and 7 Nov. 1909, Edwin H. Hall Papers, Houghton Library, Harvard University. For Franklin's relationship to Bridgman, see Moyer, "P. W. Bridgman's Operational Perspective on Physics," forthcoming.

4. Franklin, "Popular Science," pp. 362–65. See also Franklin's review of *Atoms and Energies,* by D. A. Murray, *Science,* 14 (1901), pp. 295–96.

5. Franklin, review of *Traité Élémentaire de Mechanique Chemique,* by P. Duhem, *Phys. Rev.,* 6 (1898), p. 173, n.1, and 8 (1899), p. 354.

6. Franklin, "Popular Science," p. 368.

7. Franklin and Macnutt, *A Calendar of Leading Experiments,* pp. 70–73. Franklin, "Operative versus Abstract Philosophy in Physics," pp. 623–25.

8. Franklin, "The Principle of Relativity," pp. 1–21. Franklin, "The Quantum Puzzle and Time," pp. 258–61. Franklin to Percy Bridgman, 8 Nov. 1929, Bridgman Papers, by permission of the Harvard University Archives.

[189]

17.   ADAMS: FROM UNITY TO MULTIPLICITY

1. In his annotated edition of *The Education of Henry Adams,* Ernest Samuels notes that Adams actually began writing in 1903 and continued into 1906. Although Adams privately printed copies of the book in 1906 and 1907, it was not publicly available until after his death in 1918. See Adams, *The Education of Henry Adams,* pp. xxi and 539, n.3. The following quotations from the Samuels edition of Adams's *Education* (copyright 1918 by the Massachusetts Historical Society; copyright 1946 by Charles F. Adams) are reprinted by permission of Houghton Mifflin Company.

2. Adams, *Education,* pp. 376-77. Regarding Langley's kinship to Adams, see Vaeth, *Langley,* p. 5. Regarding Langley and Adams's friendship, see Adams's comments in *Letters of Henry Adams,* pp. 11, 325, 389, 396, 420, 649.

3. Adams, *Education,* pp. 449-61. Jordy, *Henry Adams,* esp. pp. 227-36; Wasser, *The Scientific Thought of Henry Adams,* esp. pp. 116-27; and Samuels, *Henry Adams,* pp. 384-85. Brush, *The Temperature of History,* pp. 121-27.

4. Adams, *Education,* pp. 452, 455-57.

5. Adams inscribed, with the date "1903," his heavily annotated copy of *The Grammar of Science* (2d ed., 1900); see *Education,* ed. Samuels, p. 672, n. 3. Pearson, *The Grammar of Science* (1900), pp. vii-xi; see also 1st ed. (1892), pp. vii-ix.

6. Adams, *Education,* pp. 450-51.

7. Joseph Jastrow, rev. of *The Grammar of Science, Science,* 12 (1900), pp. 67-69; for a similar, positive review, see also *Pop. Sci. Mo.,* 57 (1900), p. 550. Not all the American reviews, however, were favorable. Just as he objected to Mach, Charles Peirce also objected to Pearson with his nominalistic tendencies; see Peirce, "Pearson's Grammar of Science, Annotations on the First Three Chapters," *Pop. Sci. Mo.,* 58 (1901), p. 296-306. See also Peirce's negative review of the 1st ed. in *The Nation,* 55 (1892), 15.

8. Adams, *Education,* pp. 377, 449-50.

9. Josiah Royce, "Introduction," in Poincaré, *Science and Hypothesis,* p. xvi.

10. Stallo, *Concepts and Theories,* ed. Bridgman, p. viii.

11. Adams, *Education,* pp. 450-52. For a summary of the papers Langley reprinted during 1887 to 1905, see the "Index: 1849-1961" of the Smithsonian *Annual Reports,* esp. under the topics of "Atomic Theory," "Electricity," "Light," "Matter," and "Physics."

12. [Alfred Mayer], "Modern Physics," *The Critic,* 2 (25 Feb. 1882), p. 58. Adams, *Education,* p. 452.

BIBLIOGRAPHY

The bibliography is divided into primary and secondary works that were consulted in preparing this study. Included are all major books, articles, and other writings that are cited in the shortened form of endnote. Minor items such as nineteenth-century book reviews are not included; bibliographic details on these particular items can be found in expanded endnotes.

PRIMARY SOURCES

Adams, Henry. *The Education of Henry Adams.* Edited by Ernest Samuels. 1918. Reprint. Boston: Houghton Mifflin, 1974.

———. *Letters of Henry Adams (1892–1918).* Edited by Worthington C. Ford. Boston: Houghton Mifflin, 1938.

Barker, George F. *The Correlation of Vital and Physical Forces.* University Series, no. 2. New Haven, Conn.: Chatfield, 1870.

Boltzmann, Ludwig. "Reise eines deutschen Professors ins Eldorado." In *Populäre Schriften.* Leipzig: Barth, 1905.

Bridgman, Percy W. *Dimensional Analysis.* Rev. ed. New Haven: Yale University Press, 1931.

———. *The Logic of Modern Physics.* New York: Macmillan, 1927.

———. Percy Bridgman Papers. Harvard University Archives, Cambridge, Mass. (HUG 4234.8)

———. *Reflections of a Physicist.* New York: Philosophical Library, 1950.

Congress of Arts and Science: Universal Exposition, St. Louis, 1904. Vol. 1, *Philosophy and Mathematics.* Edited by Howard J. Rogers. Boston: Houghton & Mifflin, 1905.

Congress of Arts and Science: Universal Exposition, St. Louis, 1904. Vol. 4, *Inorganic Sciences.* Edited by Howard J. Rogers. Boston: Houghton & Mifflin, 1906.

Crew, Henry. Henry Crew Collection. Niels Bohr Library, Center for History of Physics, American Institute of Physics, New York.

—. "Recent Advances in the Teaching of Physics." *Science* 19 (1904): 481–88.

—, ed. *The Wave Theory of Light*. Vol. 10 of *Scientific Memoirs*. Edited by J. S. Ames. New York: American Book, 1900.

—. "What Can Be Done To Make the Study of Physics a Better 'Training for Power'?" *The School Review* 8 (1900): 520–27.

Davis, William H. "The International Congress of Arts and Science." *Popular Science Monthly* 66 (1904): 5–32.

Dolbear, Amos E. *The Art of Projecting: A Manual of Experimentation . . . with the Porte Lumiere and Magic Lantern*. 2d ed. Boston: Lee & Shepard, 1888.

—. *First Principles of Natural Philosophy*. Boston: Ginn, 1897.

—. *Matter, Ether, and Motion*. Boston: Lee & Shepard, 1892.

—. *Modes of Motion: or Mechanical Conceptions of Physical Phenomena*. Boston: Lee & Shepard, 1897.

Duhem, Pierre. *The Aim and Structure of Physical Theory*. Translated by Philip P. Wiener. 2d ed., 1914. Reprint. Princeton: Princeton University Press, 1954.

Electrical Conference, Philadelphia, 1884. *Report of the Electrical Conference at Philadelphia in September, 1884*. Washington, D.C.: GPO, 1886.

Ford, Worthington C., ed. *Letters of Henry Adams (1892–1918)*. Boston: Houghton Mifflin, 1938.

Franklin, William S. "Operative versus Abstract Philosophy in Physics." *Science* 63 (1926): 623–25.

—. "Popular Science." *Proceedings of the AAAS* 52 (1902–03): 361–68. Reprinted in *Science* 17 (1903): 8–15.

—. "The Principle of Relativity." *Journal of the Franklin Institute* (July 1911): 1–21.

—. "The Quantum Puzzle and Time." *Science* 60 (1924): 258–61.

—. William Franklin Papers. Institute Archives and Special Collections, Massachusetts Institute of Technology, Cambridge, Mass.

————, and Macnutt, Barry. *A Calender of Leading Experiments.* South Bethlehem, Pa.: Franklin, Macnutt and Charles, 1918.

Gibbs, J. Willard. *The Collected Works of J. Willard Gibbs.* 2 vols. New York: Longmans & Green, 1928.

Hall, Edwin H. Edwin H. Hall Papers. Houghton Library, Harvard University, Cambridge, Mass.

————. "Experiments on the Effect of Magnetic Force on the Equipotential Lines of an Electric Current." *American Journal of Science,* 3d Ser. 36 (1888): 131–46; 277–86.

————. "On the New Action of Magnetism on a Permanent Electric Current." *American Journal of Science,* 3d Ser. 20 (1880): 161–86.

————. "On the Rotation of the Equipotential Lines of an Electric Current by Magnetic Action." *American Journal of Science,* 3d Ser. 29 (1885): 117–35.

————. "Teaching Elementary Physics," Pt. II. *Educational Review* 5 (1893): 325–33.

————. "A Tentative Theory of Thermo-Electric Action." *Proceedings of the AAAS* 54 (1904): 373–88.

————, and Bergen, Joseph Y. *A Textbook of Physics, Largely Experimental.* Revised and enlarged. New York: Holt. 1897.

————, and Smith, Alexander. *The Teaching of Chemistry and Physics in the Secondary School.* New York: Longmans & Green, 1902. Revised ed., 1910.

Helmholtz, Hermann von. *Epistemological Writings: The Paul Hertz/ Moritz Schlick Centenary Edition of 1921.* Edited by Robert S. Cohen and Yehuda Elkana. Vol. 37 of Boston Studies in the Philosophy of Science. Dordrecht, Holland: Reidel, 1977.

————. *Popular Scientific Lectures.* Edited by Morris Kline. New York: Dover, 1962.

Holman, Silas W. *Matter, Energy, Force and Work: A Plain Presentation of Fundamental Physical Concepts and of the Vortex-Atom and Other Theories.* New York: Macmillan, 1898.

James, William. *Pragmatism: A New Name for Some Old Ways of Thinking* [1907]. Reprinted in *The American Pragmatists,* edited by Milton Konvitz and Gail Kennedy. New York: Meridian, 1960.

[193]

Langley, Samuel P. "The History of a Doctrine." *American Journal of Science,* 3d Ser. 37 (1889): 1–23.

———. "The Laws of Nature." *Science* 15 (1902): 921–27.

———. *The New Astronomy.* 1884. Reprint. Boston: Houghton & Mifflin, 1896.

Lorentz, Hendrik. *The Theory of Electrons.* New York: Columbia University Press, 1909.

Mach, Ernst. *Popular Scientific Lectures.* Translated by Thomas Mc-Cormack. 2d ed. Chicago: Open Court, 1897.

———. *The Science of Mechanics: A Critical and Historical Account of Its Development.* Translated by Thomas McCormack. 9th ed., 1933. Reprint. La Salle, Ill.: Open Court, 1942.

Mann, C. Riborg. *The Teaching of Physics for Purposes of General Education.* New York: Macmillan, 1912.

Mayer, Alfred M. *The Earth a Great Magnet.* New Haven: Chatfield, 1872.

———. "Henry as a Discoverer." *Smithsonian Miscellaneous Collections* 21 (1881): 475–508.

———. *Lecture-Notes on Physics.* Philadelphia: Franklin Institute, 1868.

———. "On the Morphological Laws of the Configurations Formed by Magnets . . . With Notes on Some of the Phenomena in Molecular Structure which these Experiments May Serve to Explain and Illustrate." *American Journal of Science and Arts,* 3d Ser. 16 (1878): 247–56.

———. "On the Physical Conditions of a Closed Circuit Contiguous to a Permanent and Constant Voltaic Current; or on 'The Electrotonic State.'" *American Journal of Science and Arts,* 3d Ser. 1 (1871): 17–24.

———. *Sound: A Series of Simple, Entertaining, and Inexpensive Experiments in the Phenomena of Sound, for the Use of Students of Every Age.* New York: Appleton, 1879.

———, and Barnard, Charles. *Light: A Series of Simple, Entertaining, and Inexpensive Experiments in the Phenomena of Light, for the Use of Students of Every Age.* New York: Appleton, 1879.

Mendenhall, Thomas C. *A Century of Electricity*. London: Macmillan, 1887.

———. "Physics." In *The Progress of the Century*. New York: Harper, 1901.

Michelson, Albert A. *Light Waves and Their Uses*. Chicago: University of Chicago Press, 1903.

———. "A Plea for Light Waves." *Proceedings of the AAAS* 37 (1888): 67–78.

———. "The Relative Motion of the Earth and the Ether." *American Journal of Science*, 4th Ser. 3 (1897): 475–78.

———. "The Relative Motion of the Earth and the Luminiferous Ether." *American Journal of Science*, 3d Ser. 22 (1881): 120–29.

———. *Studies in Optics*. Chicago: University of Chicago Press, 1927.

———. "A Theory of the X-rays." *American Journal of Science*, 4th Ser. 1 (1896):312–14.

———, and Morley, Edward. "Influence of Motion of the Medium on the Velocity of Light." *American Journal of Science*, 3d Ser. 31 (1886): 377–86.

———. "On the Relative Motion of the Earth and the Luminiferous Ether." *American Journal of Science*, 3d Ser. 34 (1887): 333–45.

Millikan, Robert A. *Autobiography*. New York: Prentice-Hall, 1950.

———, and Gale, Henry G. *A First Course in Physics*. Boston: Ginn, 1906.

Newcomb, Simon. "The Coming International Congress of Arts and Science at St. Louis, September 19–24." *Popular Science Monthly* 65 (1904): 466–73.

———. "The Course of Nature." *Popular Science Monthly Supplement*. No. 18 (Oct. 1878): 481–93.

———. "Exact Science in America." *North American Review* 119 (1874): 286–308.

———. *His Wisdom, The Defender: A Story*. 1900. Reprint. New York: Arno, 1975.

———. "Modern Scientific Materialism." *The Independent* 32 (23 Dec. 1880): 1.

[195]

————. "The Relation of Scientific Method to Social Progress" (1880). In *Side-Lights on Astronomy and Kindred Fields of Popular Science: Essays and Addresses.* New York: Harper and Bros., 1906:312–29.

————. *The Reminiscences of an Astronomer.* Boston: Houghton & Mifflin, 1903.

————. *Sidelights on Astronomy and Kindred Fields of Popular Science: Essays and Addresses.* New York: Harper, 1906.

————. Simon Newcomb Papers. Manuscript Division, Library of Congress, Washington, D. C.

————. "Speculative Science." *International Review* 12 (1882): 334–41.

Nipher, Francis E. *Electricity and Magnetism: A Mathematical Treatise for Advanced Undergraduate Students.* St. Louis: Boland, 1895.

————. "The Ether." *Proceedings of the AAAS* 40 (1891): 127–33.

————. "Physics During the Last Century." *Transactions of the St. Louis Academy of Science* 11 (1901) 105–23.

————. *Thoughts of Our Conceptions of Physical Law.* Kansas City: Kansas City Review, 1878.

Norton, William A. "On Molecular Physics." *American Journal of Science and Arts,* 2d Ser. 40 (1865): 61–73.

Page, Leigh. "A Century's Progress in Physics." In *A Century of Science in America: with Special Reference to the American Journal of Science, 1818–1918.* New Haven: Yale University Press, 1918.

Pearson, Karl. *The Grammar of Science.* London: Scott, 1892.

————. *The Grammar of Science.* 2d ed. London: Black, 1900.

Peirce, Benjamin. *Ideality in the Physical Sciences.* Boston: Little & Brown, 1881.

Peirce, Charles S. *Charles S. Peirce: Essays in the Philosophy of Science.* Edited by Vincent Tomas. New York: Liberal Arts Press, 1957.

————. *Charles S. Peirce: Selected Writings.* Edited by Philip P. Wiener. 1958. Reprint. New York: Dover, 1966.

————. *Collected Papers of Charles Sanders Peirce.* 8 vols. Edited by Arthur W. Burks. Cambridge, Mass.: Harvard University Press, 1958–63.

[196]

———. *Philosophical Writings of Peirce.* Edited by Justus Buchler. 1940. Reprint. New York: Dover, 1955.

Poincaré, Henri. *Science and Hypotheses.* Translated by George Halsted. New York: Science Press, 1905.

———. *The Value of Science.* Translated by George Halsted. New York: Science Press, 1907.

Rowland, Henry A. *The Physical Papers of Henry Augustus Rowland.* Baltimore: Johns Hopkins Press, 1902.

Stallo, John B. *The Concepts and Theories of Modern Physics.* London: Kegan Paul, Trench, 1882. Also published New York: Appleton, 1882.

———. *The Concepts and Theories of Modern Physics.* Edited by Percy W. Bridgman. 3d ed., 1888. Reprint. Cambridge: Belknap, 1960.

———. "Introduction to the Second Edition." In *The Concepts and Theories of Modern Physics,* edited by Percy W. Bridgman. 3d ed., 1888. Reprint. Cambridge, Mass.: Belknap, 1960.

———. "The Primary Concepts of Modern Physical Science." *Popular Science Monthly* 3 (1873): 705-17; 4 (1873-74): 92-108, 219-31, 349-61.

———. *Reden, Abhandlungen und Briefe.* New York: E. Steiger, 1893.

———. "Speculative Science." *Popular Science Monthly* 21 (1882): 145-64.

Stewart, Balfour, and Tait, Peter G. *The Unseen Universe: or Physical Speculations on a Future State.* 5th ed. London: Macmillan, 1886.

Thomson, Joseph John. *Electricity and Matter.* New Haven: Yale University Press, 1904.

———. *The Discharge of Electricity through Gases.* New York: Scribner's, 1898.

———. *Recollections and Reflections.* New York: Macmillan, 1937.

Thomson, William (Lord Kelvin). *Baltimore Lectures on Molecular Dynamics and the Wave Theory of Light.* London: Clay, 1904.

———. *Notes of Lectures on Molecular Dynamics and the Wave Theory of Light.* Baltimore: 1884. "Papyrograph" reproduction.

———. "The Wave Theory of Light." In *Popular Lectures and Addresses*. London: 1899. Reprinted in The Harvard Classics, edited by Charles W. Eliot. Vol. 30, *Scientific Papers*. New York: Collier, 1910.

———, and Tait, Peter G. *Treatise on Natural Philosophy*. Rev. ed. Cambridge: Cambridge University Press, 1879.

Trowbridge, John. *The New Physics: A Manual of Experimental Study for High School and Preparatory Schools for Colleges*. New York: Appleton, 1884.

———. "On the Heat Produced by the Rapid Magnetization and Demagnetization of the Magnetic Metals." *Proceedings of the American Academy of Arts and Sciences,* N.S. 6 (1878–79): 114–21.

———. "On Vortex Rings in Liquids." *Proceedings of the American Academy of Arts and Sciences,* N.S. 4 (1876–77): 131–36.

———. "Physical Science of Today." *Atlantic Monthly* 103 (1909): 318–24.

———. "What Is Electricity?" *Popular Science Monthly* 26 (1884): 76–88.

———. *What is Electricity?* 1896. Reprint. New York: Appleton, 1899.

———, and Rollins, William. "Radium and the Electron Theory." *American Journal of Science,* 4th Ser. 18 (1904): 77–79.

Tyndall, John. *Lectures on Light: Delivered in the United States in 1872–73*. New York: Appleton, 1873.

———. *Proceedings at the Farewell Banquet to Professor Tyndall: Given at Delmonico's, New York, February 4, 1873*. New York: Appleton, 1873.

Wright, Chauncey. *Letters of Chauncey Wright*. Edited by James B. Thayer. Cambridge: Wilson, 1878.

———. *Philosophical Discussions*. Edited by Charles Norton. New York: Holt, 1877.

———. *The Philosophical Writings of Chauncey Wright*. Edited by Edward Madden. New York: Liberal Arts Press, 1958.

SECONDARY SOURCES

Badash, Lawrence. "The Completeness of Nineteenth-Century Science." *Isis* 63 (1972): 48–58.

————. "The Origins of Big Science: Rutherford at McGill." In *Rutherford and Physics at the Turn of the Century*, edited by Mario A. Bunge and William R. Shea. New York: Science History, 1979.

————. *Radioactivity in America: Growth and Decay of a Science*. Baltimore: Johns Hopkins University Press, 1979.

————, ed. *Rutherford and Boltwood: Letters on Radioactivity*. New Haven: Yale University Press, 1969.

Berkson, William. *Fields of Force: The Development of a World View from Faraday to Einstein*. London: Routledge and Kegan Paul, 1974.

Bernstein, Jeremy. *Einstein*. New York: Viking, 1973.

Borut, Michael. "The Scientific American in 19th-Century America." Ph.D. diss., New York University, 1978.

Brown, Sanborn C., and Rieser, Leonard M. *Natural Philosophy at Dartmouth: From Surveyors' Chains to the Pressure of Light*. Hanover, New Hampshire: University Press of New England, 1974.

Brush, Stephen G. *The Kind of Motion We Call Heat: A History of the Kinetic Theory of Gases in the 19th Century*. Vol. 6 in *Studies in Statistical Mechanics*. 2 vols. Amsterdam: North-Holland, 1976.

————. *The Temperature of History: Phases of Science and Culture in the Nineteenth Century*. New York: Burt Franklin, 1978.

Buchwald, Jed Z. "The Hall Effect and Maxwellian Electrodynamics in the 1880s." *Centaurus* 23 (1979–80): 51–99, 118–62.

Bunge, Mario A., and Shea, William R., eds. *Rutherford and Physics at the Turn of the Century*. New York: Science History, 1979.

Burnham, John C. *Science in America: Historical Selections*. New York: Holt, Rinehart and Winston, 1971.

Cantor, G. N., and Hodge, M. J. S., eds. *Conceptions of Ether: Studies in the History of Ether Theories, 1740–1900*. Cambridge: Cambridge University Press, 1981.

Chamber's Dictionary of Scientists. New York: Dutton, 1951. Contains biography of Arnold Reinold.

Cohen, Robert S., and Seeger, Raymond J., eds. *Ernst Mach: Physicist and Philosopher.* Vol. 6 of Boston Studies in the Philosophy of Science. Dordrecht, Holland: Reidel, 1970.

Conkin, Paul. *Puritans and Pragmatists.* New York: Dodd & Mead, 1968.

Crew, Henry. *The Rise of Modern Physics.* Baltimore: Williams & Wilkins, 1928.

Daniels, George H. *American Science in the Age of Jackson.* New York: Columbia University Press, 1968.

————, ed. *Nineteenth-Century American Science: A Reappraisal.* Evanston: Northwestern University Press, 1972.

Daub, Edward E. "Gibbs' Phase Rule: A Centenary Retrospect." *Journal of Chemical Education* 53 (1975): 747–51.

Dictionary of American Biography. Edited by Dumas Malone. New York: Scribners, 1929–34. Contains biographies of Dewitt Brace and Francis Nipher.

The Dictionary of Canadian Biography. 2d ed. Toronto: Macmillan, 1945. Contains biography of W. D. LeSueur.

Dictionary of Scientific Biography. Edited by Charles C. Gillespie. New York: Scribner's, 1970–76. Contains biographies of Percy Bridgman, G. Stanley Hall, Hermann von Helmholtz, Joseph Henry, J. Willard Gibbs, Robert Millikan, and William Thomson.

Donnan, F. G., and Haas, Arthur, eds. *A Commentary on the Scientific Writings of J. Willard Gibbs.* 2 vols. New Haven: Yale University Press, 1936.

Doran, Barbara G. "Origins and Consolidation of Field Theory in Nineteenth-Century Britain: From the Mechanical to the Electromagnetic View of Nature." *Historical Studies in the Physical Sciences* 6 (1975): 133–260.

Drake, Stillman. "J. B. Stallo and the Critique of Classical Physics." In *Men and Moments in the History of Science,* edited by Herbert M. Evans. Seattle: University of Washington Press, 1959.

[200]

Dupree, A. Hunter, ed. *Science and the Emergence of Modern America 1865–1916.* Berkeley Series in American History. Chicago: Rand McNally, 1963.

Easton, Loyd D. *Hegel's First American Followers.* Athens, Ohio: Ohio University Press, 1966.

Eisele, Carolyn. "The Charles S. Peirce-Simon Newcomb Correspondence." *Proceedings of the American Philosophical Society* 101 (1957): 409–33.

Elkana, Yehuda, ed. *The Interaction Between Science and Philosophy.* Atlantic Highlands, N.J.: Humanities, 1974.

Eve, A. S., and Creasey, C. H. *Life and Work of John Tyndall.* London: Macmillan, 1945.

Everitt, C. W. F. *James Clerk Maxwell: Physicist and Natural Philosopher.* New York: Scribner's, 1975.

Forman, Paul. "Weimar Culture, Causality, and Quantum Theory, 1918–1927: Adaptation by German Physicists and Mathematicians to a Hostile Intellectual Environment." *Historical Studies in the Physical Sciences* 3 (1971): 1–115.

————; Heilbron, John L.; and Weart, Spencer. "Physics *circa* 1900: Personnel, Funding, and Productivity of the Academic Establishments." *Historical Studies in the Physical Sciences* 5 (1975): 1–128.

Garber, Elizabeth. "Molecular Science in Late Nineteenth-Century Britain." *Historical Studies in the Physical Sciences* 9 (1978): 265–97.

Goldberg, Stanley. "Early Response to Einstein's Theory of Relativity, 1905–1912: A Case Study in National Differences." Ph.D. diss., Harvard University, 1969.

Hall, Edwin H. "John Trowbridge." *Harvard Graduates' Magazine* (June 1923): 526–27.

————. "Michelson and Rowland." *Science* 73 (1931): 615.

————. "Physics." In *The Development of Harvard University,* edited by Samuel E. Morison. Cambridge: Harvard University Press, 1930.

————. "Physics Teaching at Harvard Fifty Years Ago." *American Physics Teacher* 6 (1938): 17–20.

[201]

Harman, P. M. *Energy, Force, and Matter: The Conceptual Development of Nineteenth-Century Physics.* Cambridge History of Science. Cambridge: Cambridge University Press, 1982.

Hartshorne, Charles. "Charles Peirce and Quantum Mechanics." *Transactions of the Charles S. Peirce Society* 9 (1973): 191–201.

Haskell, David C. *Indexes of Titles and Contributors to the Nation.* New York: New York Public Library, 1951.

Hass-Lorentz, G. L. de, ed. *H. A. Lorentz: Impressions of His Life and Work.* Amsterdam: North Holland, 1957.

Heilbron, John L. "J. J. Thomson and the Bohr Atom." *Physics Today* 30 (1977): 23–30.

Hiebert, Erwin N. "The Energetics Controversy and the New Thermodynamics." In *Perspectives in the History of Science and Technology,* edited by D. H. D. Roller. Norman, Okla.: University of Oklahoma Press, 1971.

———. "The State of Physics at the Turn of the Century." In *Rutherford and Physics at the Turn of the Century,* edited by Mario A. Bunge and William R. Shea. New York: Science History, 1979.

Hirosige, Tetu. "The Ether Problem, the Mechanistic Worldview, and the Origins of the Theory of Relativity." *Historical Studies in the Physical Sciences* 7 (1976): 3–82.

Holton, Gerald. "The Migration of the Physicists to the United States." In an Einstein centennial volume. Washington, D.C., Smithsonian Institution. Forthcoming.

———. *Thematic Origins of Scientific Thought: Kepler to Einstein.* Cambridge: Harvard University Press, 1973.

Jaffe, Bernard. *Michelson and the Speed of Light.* Garden City, N.Y.: Anchor, 1960.

Jordy, William H. *Henry Adams: Scientific Historian.* New Haven: Yale University Press, 1952.

Kargon, Robert H. "The Conservative Mode: Robert A. Millikan and the Twentieth Century Revolution in Physics." *Isis* 68 (1977): 509–26.

Kevles, Daniel J. *The Physicists: The History of a Scientific Community in Modern America.* New York: Knopf, 1978.

[202]

————. "The Physics, Mathematics, and Chemical Communities in America, 1870–1915: A Statistical Survey." *Social Science Working Paper,* no. 136. Pasadena: California Institute of Technology, 1977.

————. "The Physics, Mathematics, and Chemical Communities in the United States, 1870 to 1915: A Preliminary Statistical Report." *Social Science Working Paper,* no. 94. Pasadena: California Institute of Technology, 1975.

————. "The Physics, Mathematics, and Chemistry Communities: A Comparative Analysis." In *The Organization of Knowledge in Modern America, 1860–1920,* edited by Alexandra Oleson and John Voss. Baltimore: Johns Hopkins University Press, 1979.

————. "Physics, Mathematics, and Chemistry in America, 1870–1915: A Comparative Institutional Analysis." *Social Science Working Paper,* no. 139. Pasadena, California Institute of Technology, 1976.

————. "Robert A. Millikan." *Scientific American* 240 (1979): 142–51.

————. "The Study of Physics in America, 1865–1916." Ph.D. diss., Princeton University, 1964.

————; Sturchio, Jeffrey L.; and Carroll, P. Thomas. "The Sciences in America, Circa 1880." *Science* 209 (1980): 27–32.

Klein, Martin J. "The Early Papers of J. Willard Gibbs: A Transformation of Thermodynamics." *Proceedings of the 15th International Congress of the History of Science* (1978): 330–41.

————. "Gibbs on Clausius." *Historical Studies in the Physical Sciences* 1 (1969): 127–49.

————. "Maxwell, His Demon and the Second Law of Thermodynamics." *American Scientist* 58 (1970): 84–97.

————. "Mechanical Explanation at the End of the Nineteenth Century." *Centaurus* 17 (1973): 58–82.

————. *Paul Ehrenfest: The Making of a Theoretical Physicist.* New York: American Elsevier, 1970.

Knowlton, A. A. "Henry Crew." *Isis* 45 (1954): 168–74.

Kohlstedt, Sally G. *The Formation of the American Scientific Community: The American Association for the Advancement of Science, 1848–60.* Urbana: University of Illinois Press, 1976.

[203]

Konvitz, Milton R., and Kennedy, Gail, eds. *The American Pragmatists.* New York: Meridian, 1960.

Kuhn, Thomas S. *Black-Body Theory and the Quantum Discontinuity.* New York: Oxford University Press, 1978.

————. *The Essential Tension: Selected Studies in Scientific Tradition and Change.* Chicago: University of Chicago Press, 1979.

————. *The Structure of Scientific Revolutions.* 2d ed., enl. Chicago: University of Chicago Press, 1970.

Lenzen, Victor F. *Benjamin Peirce and the U.S. Coast Survey.* San Francisco: San Francisco Press, 1968.

————. "Charles S. Peirce as Mathematical Physicist." *Transactions of the Charles S. Peirce Society* 11 (1975): 159–66.

Leverette, William E. "Science and Values: A Study of Edward L. Youmans' *Popular Science Monthly, 1872–1887.*" Ph.D. diss., Vanderbilt University, 1963.

Livingston, Dorothy Michelson. *The Master of Light: A Biography of Albert A. Michelson.* New York: Scribner's, 1973.

Lyman, Theodore. "Recollections [on the Jefferson Laboratory]." University Archives, Harvard University Library, Cambridge, Mass. (Jan. 1944) HUG 4540.2. Typescript.

McCormack, John. "John Bernard Stallo." *The Open Court* 14 (1900): 276–83.

McCormmach, Russell. "H. A. Lorentz and the Electromagnetic View of Nature." *Isis* 61 (1970): 459–97.

Madden, Edward H. *Chauncey Wright and the Foundations of Pragmatism.* Seattle: University of Washington Press, 1963.

Miller, John D. "Rowland and the Nature of Electric Currents." *Isis* 63 (1972): 5–27.

————. "Rowland's Magnetic Analogy to Ohm's Law." *Isis* 66 (1975): 230–41.

————. "Rowland's Physics." *Physics Today* 29, no. 7 (1976): 39–45.

Molella, Arthur P., and Reingold, Nathan. "Theorists and Ingenious Mechanics: Joseph Henry Defines Science." *Science Studies* (London) 3 (1973): 323–51.

————; Reingold, Nathan; Rothenberg, Marc; Steiner, John F.; and Waldenfels, Kathleen, eds. *A Scientist in American Life: Essays and Lectures of Joseph Henry.* Washington, D.C.:Smithsonian Institution, 1980.

Morison, Samuel E., ed. *The Development of Harvard University.* Cambridge: Harvard University Press, 1930.

Mott, Frank L. *A History of American Magazines, 1885–1905.* Cambridge: Harvard University Press, 1957.

Moyer, Albert E. "In Advocacy of Science: The Essays of Simon Newcomb." Forthcoming.

————. "P. W. Bridgman's Operational Perspective on Physics: Origins, Development, and Reception." Forthcoming.

————. "Edwin Hall and the Emergence of the Laboratory in Teaching Physics." *Physics Teacher* 14 (1976): 96–103.

Moyer, Donald F. "Continuum Mechanics and Field Theory: Thomson and Maxwell." *Studies in the History and Philosophy of Science* 9 (1978): 35–50.

————. "Energy, Dynamics, Hidden Machinery: Rankine, Thomson and Tait, Maxwell." *Studies in the History and Philosophy of Science* 8 (1977): 251–68.

————. "The Use of Dynamics as the Basis of Physical Theory by British Theoretical Physicists in the Latter Half of the Nineteenth Century." Ph.D. diss., University of Wisconsin-Madison, 1973.

Murphey, Murray G. *The Development of Peirce's Philosophy.* Cambridge: Harvard University Press, 1961.

National Academy of Sciences. Biographical Memoirs. Washington: NAS, 1909–59. Contains biographies of Carl Barus, J. Willard Gibbs, Edwin Hall, Alfred Mayer, Thomas Mendenhall, Robert Millikan, Simon Newcomb, Edward Nichols, and John Trowbridge.

The National Cyclopaedia of American Biography. New York: James T. White, 1907–32. Contains biographies of Amos Dolbear, Robert Eccles, William Franklin, Francis Nipher, and William Norton.

Norberg, Arthur L. "Simon Newcomb and Nineteenth-Century Positional Astronomy." Ph.D. diss., University of Wisconsin-Madison, 1974.

[205]

———. "Simon Newcomb's Early Astronomical Career." *Isis* 69 (1978): 209–25.

Numbers, Ronald L. "The Making of an Eclectic Physician: Joseph M. McElhinney and the Eclectic Medical Institute of Cincinnati." *Bulletin of the History of Medicine* 47 (1973): 155–66.

Obendorf, Donald L. "Samuel P. Langley: Solar Scientist, 1867–1891." Ph.D. diss., University of California, Berkeley, 1969.

Olson, Richard. *Scottish Philosophy and British Physics, 1750–1880: A Study in the Foundations of the Victorian Scientific Style.* Princeton: Princeton University Press, 1975.

Paradis, James, and Postlewait, Thomas, eds. *Victorian Science and Victorian Values: Literary Perspectives.* Vol. 360 of *Annals of the New York Academy of Sciences.* New York: New York Academy of Sciences, 1981.

Peterson, Sven. "Benjamin Peirce: Mathematician and Philosopher." In *Roots of Scientific Thought: A Cultural Perspective,* edited by Philip P. Weiner and Aaron Noland. New York: Basic Books, 1957.

Philips, Melba, ed. *AAPT Pathways: Proceedings of the Fiftieth Anniversary Symposium of the AAPT.* Stony Brook, N.Y.: American Association of Physics Teachers, 1981.

———. *On Teaching Physics: Reprints of American Journal of Physics Articles from the First Half Century of AAPT.* Stony Brook, N.Y.: American Association of Physics Teachers, 1979.

Plotkin, Howard. "Edward C. Pickering, the Henry Draper Memorial, and the Beginning of Astrophysics in America." *Annals of Science* 35 (1978): 365–77.

Poole's Index to Periodical Literature. 1888. Reprint. New York: Smith, 1938.

Pyenson, Lewis. "The Incomplete Transmission of a European Image: Physics at Greater Buenos Aires and Montreal, 1890–1920." *Proceedings of the American Philosophical Society* 122 (1978): 92–114.

Reilly, Francis E. *Charles Peirce's Theory of Scientific Method.* New York: Fordham University Press, 1970.

Reingold, Nathan, ed. *Science in America Since 1820.* New York: Science History Pub., 1976.

[206]

———, ed. *Science in Nineteenth-Century America: A Documentary History.* New York: Hill and Wang, 1964.

———, ed. *The Sciences in the American Context: New Perspectives.* Washington D.C.: Smithsonian Institution, 1979.

Rosen, Sidney. "A History of the Physics Laboratory in the American Public High School (to 1910)." *American Journal of Physics* 22 (1954): 194–204.

Rosensohn, William L. *The Phenomenology of Charles S. Peirce.* Amsterdam: Grüner, 1974.

Rukeyser, Murial. *Willard Gibbs.* Garden City, N.Y.: Doubleday, 1942.

Samuels, Ernest. *Henry Adams: The Major Phase.* Cambridge, Mass.: Belknap, 1964.

Seeger, Raymond J. *J. Willard Gibbs: American Mathematical Physicist Par Excellence.* New York: Pergamon, 1974.

Servos, John W. "A Disciplinary Program That Failed: Wilder D. Bancroft and the *Journal of Physical Chemistry,* 1896–1933." *Isis* 73 (1982): 207–32.

Schaffner, Kenneth F. *Nineteenth-Century Aether Theories.* New York: Pergamon, 1972.

Shankland, R. S. "Michelson: America's First Nobel Prize Winner in Science." *The Physics Teacher* 15 (1977): 19–25.

Sharlin, Harold I. *The Convergent Century: The Unification of Science in the Nineteenth Century.* Vol. 46 in The Life of Science Library. New York: Abelard-Schuman, 1966.

———. *Lord Kelvin: Dynamic Victorian.* University Park: Pennsylvania State University Press, 1979.

Siegel, Daniel M. "Balfour Stewart and Gustav Robert Kirchhoff: Two Independent Approaches to 'Kirchhoff's Radiation Law.'" *Isis* 67 (1976): 565–600.

———. "Completeness as a Goal in Maxwell's Electromagnetic Theory." *Isis* 66 (1975): 361–68.

Silliman, Robert H. "William Thomson: Smoke Rings and Nineteenth-Century Atomism." *Isis* 54 (1963): 461–74.

[207]

Smithsonian Annual Reports: 1849–1961. Author-Subject Index to Articles. Compiled by Ruth M. Stemple. Washington, D.C.: Smithsonian Institution, 1963.

Snelders, H. A. M. "A. M. Mayer's Experiments with Floating Magnets and Their Use in the Atomic Theories of Matter." *Annals of Science* 33 (1976): 67–80.

Snow, C. P. *The Physicists.* Boston: Little, Brown, 1981.

Sopka, Katherine R. "An Apostle of Science Visits America." *The Physics Teacher* 10 (1972): 369–75.

————. "The Discovery of the Hall Effect: Edwin Hall's Hitherto Unpublished Account." In *The Hall Effect and Its Applications,* edited by C. L. Chien and C. R. Westgate. New York: Plenum, 1980.

————. "John Tyndall: International Popularizer of Science." In *John Tyndall: Essays on a Natural Philosopher,* edited by W. H. Brock, N. D. McMillan, and R. C. Mollan. Dublin: Royal Dublin Society, 1981.

————. *Quantum Physics in America, 1920–1935.* New York: Arno, 1980.

Strutt, Robert J. (Lord Rayleigh). *Life of John William Strutt, Third Baron Rayleigh.* 1924. Reprint. Madison: University of Wisconsin Press, 1968.

————. *The Life of Sir J. J. Thomson.* Cambridge: Cambridge University Press, 1943.

Swenson, Loyd S., Jr. *The Ethereal Aether: A History of the Michelson-Morley-Miller Aether-Drift Experiments, 1880–1930.* Austin: University of Texas Press, 1972.

————. *Genesis of Relativity: Einstein in Context.* Vol. 5 in Studies in the History of Science. New York: Burt Franklin, 1979.

Thayer, H. S. *Meaning and Action: A Critical History of Pragmatism.* New York: Bobbs-Merrill, 1968.

Thompson, Silvanus P. *The Life of William Thomson.* 2 vols. London: Macmillan, 1910.

Vaeth, J. Gordon. *Langley: Man of Science and Flight.* New York: Ronald, 1966.

Walker, Don D. "The Popular Science Montly, 1872–1878: A Study in the Dissemination of Scientific Ideas in America." Ph.D. diss., University of Minnesota, 1956.

Warnow, Joan N. *National Catalog of Sources for History of Physics: Report No. 1.* New York: American Institute of Physics, 1969.

Wasser, Henry H. *The Scientific Thought of Henry Adams.* Thessaloniki, Greece: 1956. Privately printed.

Weart, Spencer R., ed. *Selected Papers of Great American Physicists.* New York: American Institute of Physics, 1976.

Wheeler, Lynde P., *Josiah Willard Gibbs: The History of a Great Mind.* Rev. ed. New Haven: Yale University Press, 1952.

White, Andrew D.; Pickering, E. C.; and Chanute, Octave. "Samuel Pierpont Langley." *Smithsonian Miscellaneous Collection* 49, no. 1720 (1907).

Wiener, Philip P. *Evolution and the Founders of Pragmatism.* Cambridge: Harvard University Press, 1949.

———. "The Peirce-Langley Correspondence and Peirce's Manuscript on Hume and the Laws of Nature." *Proceedings of the American Philosophical Society* 91 (1947): 201–28.

———, and Young, Frederic H., eds. *Studies in the Philosophy of Charles Sanders Peirce.* Cambridge: Harvard University Press, 1952.

Wilkinson, George D. "John B. Stallo's Criticism of Physical Science." Ph.D. diss., Columbia University, 1951.

Williams, L. Pearce. *Album of Science: The Nineteenth Century.* New York: Scribner's Sons, 1978.

———. *Michael Faraday.* New York: Simon and Schuster, 1971.

Williams, S. R. "A History of Physics in Oberlin College." *Oberlin Alumni Magazine* (Nov. 1914): 50–56.

Wilson, David B. "Concepts of Physical Nature: John Herschel to Karl Pearson." In *Nature and the Victorian Imagination,* edited by U. C. Knoepflmacher and G. B. Tennyson. Berkeley: University of California Press, 1977.

Wilson, Edwin B. "William Suddards Franklin." *Proceedings of the American Academy of Arts and Sciences* 75 (1944): 162–64.

[209]

Wilson, Leonard G., ed. *Benjamin Silliman and his Circle: Studies on the Influence of Benjamin Silliman on Science in America.* New York: Science History, 1979.

Winnik, Herbert C. "The Role of Personality in the Science and the Social Attitudes of Five American Men of Science, 1876–1916." Ph.D. diss., University of Wisconsin, 1968.

INDEX

INDEX

AAAS, xvii, 53, 83, 88, 107, 109, 112, 137, 153, 163; Philadelphia meeting (1884) of, 46, 48, 66, 73, 75, 103
Abraham, Max, 123, 155
Adams, Henry, xviii, 107, 167–72; *Education of Henry Adams, The,* 167
Allegheny Observatory, 107
American Academy of Arts and Sciences, 99; *Proceedings* of the, 47; and Rumford Medal, 63, 99, 100
American Association of Physics Teachers, 163
American Journal of Science and Arts (later, *American Journal of Science*), 12, 41, 47, 52, 100, 104, 125
American Physical Society, xviii, 61, 125, 143, 147–48, 153, 160; *Bulletin* of the, 125
American Telephone and Telegraph, xix
Ames, Joseph: and Thomson, 121–22
Amherst College, 153

BAAS, 66; Montreal meeting (1884) of, 46–47, 52, 62, 73
Babbage, Charles, 139
Bacon, Francis, 40
Badash, Lawrence, xx, 186 n.2
Baltimore Lectures. *See* Thomson, William, and Baltimore Lectures (1884)
Barker, George, 28, 46, 52
Barus, Carl, 120, 122, 153, 155, 159, 161
Becquerel, Henri, 122, 127
Bell, Alexander Graham: and Michelson, 61, 62

Bohr, Niels, xvii, 166
Boltwood, Bertram, 148; and Rutherford, 147
Boltzmann, Ludwig, 9, 101–02; in U.S., 143, 144, 150, 153
Boscovich, Roger, 39
Boston High School and Latin School, 106
Bowdoin College, 63
Brace, Dewitt, 122, 153, 155, 157
Bridgman, Percy, 24, 77 n.19, 152, 160, 163; and Hall, 65; *Logic of Modern Physics, The,* 163; and Newcomb, 24; and Stallo, 32 n.10, 93
Brown University, 120, 153
Brush, Stephen, xvii
Buckingham, Edgar, 149, 160
Butler, Nicholas Murray, 142

Cambridge University, 19
Carnegie Institution, xix, 155
Carnot, Sadi, 100
Carus, Paul, 151
Case School of Applied Science, 59
Cauchy, Augustin, 12
Cavendish Laboratory, 147, 151
Centennial Exhibition (Philadelphia, 1876), 73
Challis, James, 11
Clark University, 59
Clausius, Rudolf, 9, 100–01,102, 103, 105
Clifford, William, 152, 163
Collège de France, 146
Columbia University, xix, 149, 151, 162
Comte, Auguste, 19–20, 26, 85, 86; *Philosophie Positive,* 20
Cooke, Josiah P., 17, 26
Cornell University, xix, 153, 162

[213]

Crew, Henry, 47, 122, 153, 155, 156, 159–60, 161; and Mach, 152; and Thomson, 75, 78
Crookes, William, 171
Curie, Marie, 122, 171–72
Curie, Pierre, 122, 146, 172

Darwin, Charles, 85, 86, 90, 92
Davis, William Harper, 142
de la Rive, Auguste, 41
Dewar, James: in U.S., 46
Dewey, John, 163
Dolbear, Amos, 42–45, 46, 47, 59, 68, 114, 116–17, 134, 136, 139; *First Principles of Natural Philosophy*, 136; *Matter, Ether, and Motion*, 44; and Mayer, 35–36; *Modes of Motion*, 136
Drake, Stillman: and Stallo, 32 n.10
Du Bois-Reymond, Emil, 6, 10, 28, 45
Duhem, Pierre, 165; and Thomson, 77–78, 79

Eccles, Robert G.: and Stallo, 18, 22
Eclectic Medical Institute of Cincinnati, 5
Einstein, Albert, xvii–xviii, xx, 61, 145, 151, 157, 166, 172
Eliot, Charles: and Hall, 66

Faraday, Michael, 39, 41, 50, 52, 56, 124; impact in U.S. of, 68, 70
Fitzgerald, Francis: in U.S., 46
Fizeau, Armand, 62
Forbes, George: in U.S., 46
Forman, Paul, xix, 176 n.1
Franklin Institute (Philadelphia), 73
Franklin, William, 162–66
Fresnel, Augustin, 12, 61–63, 77

Gage, Alfred: and Hall, 71
General Electric, xix
Gibbs, J. Willard, 46, 83, 97–105, 112, 114, 116, 123, 130–33, 134, 135, 150, 153, 155; and Adams,

167; *Elementary Principles in Statistical Mechanics*, 103, 132–33
Gilman, Daniel C.: and Gibbs, 99; and Rowland, 52; and Thomson, 75
Goldberg, Stanley, xx
Göttingen University, 68
Graham, Thomas, 28
Green, George, 55

Hale, George Ellery: and Michelson, 59
Hall, Edwin, 59, 63–67, 68, 69, 114, 116, 134, 136–37, 139, 163; and colleagues, 70–71; and Rowland, 52–53; and Trowbridge, 47
Hall, G. Stanley: and Stallo, 18, 19, 21–22, 24–25, 29, 116
Harman, P.M., xviii
Harris, William T.: and Stallo, 6
Harvard University, xix, 19, 21, 26, 47, 65, 66, 67, 68, 86, 137, 162; Jefferson Physical Laboratory, xix, 47; Lawrence Scientific School, 47, 85, 86; Observatory, 106
Hastings, Charles, 104; and Hall, 71
Heilbron, John, xix
Heisenberg, Werner, xvii, xix–xx, 166
Helm, Georg, 149; and Gibbs, 101
Helmholtz, Hermann von, 10, 24, 28, 48, 55; and Franklin, 162; and Hall, 65; impact of, in U.S., 68–69, 70; and Michelson, 59; and Rowland, 52; in U.S., 72
Henry, Joseph: and Mayer, 36, 40–41, 83; and Newcomb, 83; and Tyndall, 72
Hertz, Heinrich, 57–58, 105
History of Science Society, 160
Holman, Silas: *Matter, Energy, Force and Work*, 131; and Peirce, 131
Holton, Gerald, xvii
Hunt, E. B., 12

International Commission of Electricians (1882), 50
International Electrical Congress (St. Louis, 1904), 143
International Electrical Exposition (Philadelphia, 1884), 46
International Scientific Series (Appleton), 17, 126, 127

James, William, 47, 163, 165; and Ostwald, 148; and Peirce, 87, 131; and Wright, 87
Jevons, William S., 110
Johns Hopkins University, The, xix, 19, 21, 46, 52, 53, 65, 68, 73–75, 78, 86, 99, 121, 153
Joule, James P., 48

Kelvin, Lord. See Thomson, William
Kevles, Daniel, xviii–xx, 68, 69
Kimball, Arthur, 122, 153, 155, 157, 159, 161
Kirchoff, Gustav, 30, 68
Klein, Martin, xvii, 100
Kroenig, August, 9
Kuhn, Thomas, xvii, 176 n.1

Lagrange, Joseph, 55
Langevin, Paul, 121, 147; in U.S., 143, 146–147, 153
Langley, Samuel, 82, 83, 106–07, 109–110, 111–112, 113, 114, 116, 123, 125, 127–29, 131, 135; and Adams, 167, 169, 171; New Astronomy, The, 109, 112
Larmor, Joseph, 147, 155
Lawrence Scientific School. See Harvard University, Lawrence Scientific School
Lehigh University, 37, 162
Leiden University, 151
Le Sage, G. L., 11
LeSueur, W. D.: and Stallo, 18 n.5, 20 n.11
Lewis, Gilbert, 166
Lockyer, J. Norman, 17
Lodge, Oliver: in U.S., 46
Lorentz, Hendrik, 62, 63, 122, 124,

135, 147, 155, 166; in U.S., 151
Lovejoy, Arthur, 142; and Poincaré, 145
Lovering, Joseph, 47, 137; and Michelson, 63 n.7
Lowell Institute, 135
Ludwig, Carl, 28

MacAlister, Donald: and Stallo, 18, 19, 20–21, 22–25, 28, 31–32
McCormack, Thomas, 151
McCormmach, Russell, xvii
McGill University, 147
Mach, Ernst, 127, 161, 163, 169; Analysis of the Sensations, 151; impact in U.S. of, 151–52; Popular Scientific Lectures, 151; Science of Mechanics, 151–52; and Stallo, 93, 171
Magie, William: and Mach, 152
Massachusetts Institute of Technology, 47, 131, 162, 166
Maxwell, James Clerk, 9–14 passim, 19, 24, 26, 28, 49, 55–58 passim, 61, 65, 69, 77, 97, 99–104 passim, 112, 124, 159, 160, 169; and Gibbs, 99; impact in U.S. of, 68, 70–71; Matter and Motion, 71; and Rowland, 52, 78; Theory of Heat, 71; Treatise on Electricity and Magnetism, 55, 66, 104
Mayer, Alfred, 35–43, 46, 47, 59, 66, 68, 92, 106, 114, 115–17, 134, 137, 139, 153, 172; and Dolbear, 44, 45; and Faraday, 70; Lecture-Notes on Physics, 37; Light: Simple, Entertaining, and Inexpensive Experiments, 40; and Stallo, 18–19, 21, 23, 26, 31–32, 113; and Tyndall, 72, 73
Mendenhall, Thomas: and Hall, 71; and Thomson, 75
Michelson, Albert, 47, 52, 57, 59–63, 67, 68, 85, 112–13, 114, 116, 118, 134–36, 139, 155; and Gibbs, 99; and Thomson, 75
Mill, John Stuart, 5, 16, 40, 86, 90; and Newcomb, 85

Millikan, Robert, 122, 148, 149–50, 153, 154; and Michelson, 52, 61
Morley, Edward, 112–13; and Michelson, 59, 62–63; and Thomson, 75
Münsterberg, Hugo, 142–43

Nation, 19, 142
National Academy of Sciences, 2, 99
National Bureau of Standards, xix
National Electrical Conference (Philadelphia, 1884), 46, 73, 75, 78, 99, 103, 121; and Rowland, 55–56, 58
Nature, 42, 69
Nautical Almanac Office, 83, 85
Naval Academy, U.S., 59, 107
Naval Observatory, U.S., 85
Newcomb, Simon, 46, 52, 83–86, 88–96, 97, 114, 117–18, 131, 134, 139, 140–41, 142, 149, 155, 163; and Adams, 167; and Michelson, 59, 85; and Stallo, 18, 22, 24–25, 26–27, 29, 31–32
Newton, Isaac, 10–11, 39, 50, 110, 145, 160, 169
Nichols, Edward, 122, 149, 152, 153, 155, 157, 158, 160–61; and Franklin, 162; and Rowland, 124 n.4
Nipher, Francis, 83, 106–07, 108, 110–11, 112–13, 114, 116, 124, 134, 138–40, 155
Nobel Prize, 61, 134
Northwestern University, 153
Norton, William A., 41

Oberlin College, 47
Ohio Wesleyan University, 43
Open Court Company, 151–52
Ostwald, Wilhelm, 122, 152, 160, 163, 169; and Gibbs, 99, 101; in U.S., 143, 144, 148–50, 153; *Textbook of General Chemistry*, 149

Pearson, Karl, 127, 152, 163, 169–70; *Grammar of Science*, 169–70
Peirce, Benjamin: and Newcomb, 85–86
Peirce, Charles, 117–18, 128–29, 131, 134, 142, 149, 163, 165; and Langley, 107, 128–29; and Mach, 152; and Newcomb, 85–86, 87–88, 89, 90, 92, 93, 94; and Pearson, 170 n.7
Pennsylvania College of Gettysburg, 37
Philosophical Magazine, 47, 52, 69
Physical Review, xviii, 121, 127, 136, 148, 152, 153
Pickering, Charles, 52
Planck, Max, xvii–xviii, 133, 150, 151, 157, 172
Poincaré, Henri, 121, 152, 159, 160, 163, 169; in U.S., 143–46, 153; *Science and Hypothesis*, 171
Poisson, Simeon D., 55
Popular Science Monthly, 5, 6, 11, 14, 18, 20, 26–27, 88, 91, 94, 115, 131, 138, 140, 142, 144
Princeton University, xix, 151
Prout, William, 9
Pupin, Michael: and Mach, 152; and Ostwald, 149; and Rutherford, 148

Rayleigh, Lord, 121; and Gibbs, 103; in U.S., 47; and Michelson, 62; and Rowland, 52; and Thomson, 78
Regnault, Henri, 36
Reinold, Arnold W.: and Stallo, 18, 22
Remson, Ira, 17 n.7
Rensselaer Polytechnic Institute, 52
Roentgen, Wilhelm, 122, 123, 124, 171–72
Rogers, Howard J., 142
Rowland, Henry, 50–58, 59, 68, 114, 116–17, 118, 121, 123, 124–25, 127, 131, 134, 135, 153, 155; and Ames, 121; and colleagues, 46–47; and Faraday, 70;

[217]

$33.50